# Group Theory for Physicists

Group theory helps readers in understanding energy spectrum and the degeneracy of systems possessing discrete symmetry and continuous symmetry. This text covers two essential aspects of group theory, namely discrete groups and Lie groups. Important concepts including permutation groups, point groups, and irreducible representation related to discrete groups are discussed with the aid of solved problems. Topics such as the matrix exponential, the circle group, tensor products, angular momentum algebra, and the Lorentz group are explained to help readers in understanding the quark model and the corresponding bound states of quarks. Real life applications including molecular vibration, level splitting perturbation, and the orthogonal group are also covered. Application-oriented solved problems and exercises are interspersed throughout the text to reinforce understanding of the key concepts.

**Pichai Ramadevi** is Professor of Physics at the Indian Institute of Technology Bombay, India. She has taught courses including general relativity, group theory methods, particle physics, quantum mechanics and quantum physics. She has published more than 50 papers in journals of national and international repute. Her research interests include string theory, knot theory, quantum field theory, and supersymmetry.

**Varun Dubey** is a researcher at the International Center for Theoretical Sciences, Bangalore, India. His research interests include probability theory, stochastics in quantum dynamics, and geometric quantization.

# Group Theory for Physicists

*with Applications*

Pichai Ramadevi
Varun Dubey

# CAMBRIDGE
UNIVERSITY PRESS

University Printing House, Cambridge CB2 8BS, United Kingdom

One Liberty Plaza, 20th Floor, New York, NY 10006, USA

477 Williamstown Road, Port Melbourne, VIC 3207, Australia

314 to 321, 3rd Floor, Plot No.3, Splendor Forum, Jasola District Centre, New Delhi 110025, India

79 Anson Road, #06–04/06, Singapore 079906

Cambridge University Press is part of the University of Cambridge.

It furthers the University's mission by disseminating knowledge in the pursuit of
education, learning and research at the highest international levels of excellence.

www.cambridge.org
Information on this title: www.cambridge.org/9781108429474

© Pichai Ramadevi and Varun Dubey 2019

This publication is in copyright. Subject to statutory exception
and to the provisions of relevant collective licensing agreements,
no reproduction of any part may take place without the written
permission of Cambridge University Press.

First published 2019

Printed in India by Nutech Print Services, New Delhi 110020

*A catalogue record for this publication is available from the British Library*

*Library of Congress Cataloging-in-Publication Data*

Names: Ramadevi, Pichai, author. | Dubey, Varun, author.
Title: Group theory for physicists : with applications / Pichai Ramadevi,
Varun Dubey.
Description: Cambridge ; New York, NY : Cambridge University Press, 2019. |
Includes bibliographical references and index.
Identifiers: LCCN 2019016363 | ISBN 9781108429474 (alk. paper)
Subjects: LCSH: Group theory–Textbooks. | Mathematical physics–Textbooks.
Classification: LCC QC20.7.G76 R354 2019 | DDC 512/.2–dc23
LC record available at https://lccn.loc.gov/2019016363

ISBN 978-1-108-42947-4 Hardback
ISBN 978-1-108-45427-8 Paperback

Cambridge University Press has no responsibility for the persistence or accuracy
of URLs for external or third-party internet websites referred to in this publication,
and does not guarantee that any content on such websites is, or will remain,
accurate or appropriate.

# Contents

*List of Figures*   *ix*
*List of Tables*   *xi*
*Preface*   *xiii*

**1 Introduction**   **1**
    1.1 Definition of a Group   1
    1.2 Subgroups   4
    1.3 Conjugacy Classes   5
    1.4 Further Examples of Groups   6
    1.5 Homomorphism of Groups   8
    1.6 The Symmetric Group   9
    1.7 Direct and Semi-direct Products   14
    Exercises   16

**2 Molecular Symmetry**   **18**
    2.1 Elements of Molecular Symmetry   18
    2.2 The Symmetry Group of a Molecule   20
    2.3 Symmetry Point Groups   23
    Exercises   28

**3 Representations of Finite Groups**   **29**
    3.1 Vector Spaces   29
    3.2 Group Action on Vector Spaces   36
    3.3 Reducible and Irreducible Representations   38
    3.4 Irreducible Representations of Point Groups   41
    3.5 The Regular Representation   44
    3.6 Tensor Product of Representations   45

| | | |
|---|---|---|
| 3.7 | Decomposition of Reducible Representations | 46 |
| 3.8 | Irreducible Representations of Direct Products | 52 |
| 3.9 | Induced Representations | 54 |
| | Exercises | 57 |

## 4 Elementary Applications 59

| | | |
|---|---|---|
| 4.1 | General Considerations | 59 |
| 4.2 | Level Splitting under Perturbation | 62 |
| 4.3 | Selection Rules | 64 |
| 4.4 | Molecular Vibrations | 71 |
| | Exercises | 85 |

## 5 Lie Groups and Lie Algebras 87

| | | |
|---|---|---|
| 5.1 | The Circle Group U(1) | 87 |
| 5.2 | The Matrix Exponential | 89 |
| | 5.2.1 Generators of the Lie group | 89 |
| | 5.2.2 Convergence property of matrix exponentials | 90 |
| 5.3 | Finite Dimensional Lie Algebras | 92 |
| | 5.3.1 $\mathfrak{su}(2)$ algebra | 99 |
| 5.4 | Semi-simple Lie Algebras | 101 |
| 5.5 | Lie Algebra of a Lie Group | 103 |
| 5.6 | Examples of Lie Groups | 105 |
| 5.7 | Compact Simple Lie Algebras | 107 |
| | 5.7.1 $\mathfrak{su}(3)$ algebra | 110 |
| | 5.7.2 Cartan matrix | 113 |
| | 5.7.3 Fundamental weights | 113 |
| | Exercises | 114 |

## 6 Further Applications 115

| | | |
|---|---|---|
| 6.1 | Continuous Symmetry and Constant of Motion | 117 |
| | 6.1.1 Translational symmetry in three-dimensional space | 117 |
| 6.2 | Tensor Product Rule for $SU(2)$ Irreducible Representations | 119 |
| | 6.2.1 Digression on the Young tableau approach | 119 |
| | 6.2.2 $SU(2)$ Clebsch–Gordan matrix | 124 |
| | 6.2.3 Wigner–Eckart theorem | 128 |
| 6.3 | Elementary Particles in Nuclear and Particle Physics | 130 |
| | 6.3.1 Isospin symmetry | 131 |
| 6.4 | Quark Model | 132 |
| | 6.4.1 $SU(3)$ group approach quark model | 133 |
| | 6.4.2 Antiparticles | 136 |
| | 6.4.3 Symmetry breaking from $SU(3) \to SU(2)$ | 137 |

6.5 Non-compact Groups ... 139
    6.5.1 The Lorentz group ... 139
    6.5.2 Poincare group ... 143
    6.5.3 Scale invariance ... 143
    6.5.4 Conformal group ... 143
6.6 Dynamical Symmetry in Hydrogen Atom ... 144
    6.6.1 Lie algebra symmetry ... 145
    6.6.2 Energy levels of hydrogen atom ... 146
    Exercises ... 148

*Appendix A: Maschke's Theorem* ... *151*
*Appendix B: Schur's Lemma* ... *153*
*Bibliography* ... *156*
*Index* ... *157*

# Figures

2.1.1 Roto-Reflection Symmetry. The figure shows the action of $C_n\sigma_h$ and $\sigma_h C_n$ on an atom $A$ of some molecule . . . . . . . . . . . . 20
2.1.2 Inversion Symmetry. The central dark sphere is the point of symmetry of this hypothetical molecule . . . . . . . . . . . . 20
2.2.1 Equivalent and Non-equivalent Reflection Planes . . . . . . . . . . . . 22
2.2.2 Symmetry Axes of a Cube . . . . . . . . . . . . 22
2.2.3 Bilateral Axis . . . . . . . . . . . . 23
2.3.1 Stereographic Projections . . . . . . . . . . . . 24
2.3.2 $\sigma_d U_2$ Transformation in $D_{nd}$ Group . . . . . . . . . . . . 26
2.3.3 Symmetry Axes of the Tetrahedron . . . . . . . . . . . . 27
2.3.4 Hypothetical Molecule . . . . . . . . . . . . 28
3.1.1 $R^2$ . . . . . . . . . . . . 35
4.4.1 Normal Modes (Non-linear Tri-atomic Molecule) . . . . . . . . . . . . 75
4.4.2 Hypothetical Equilateral Molecule . . . . . . . . . . . . 78
5.1.1 The Circle Group U(1) . . . . . . . . . . . . 88
5.3.1 Group Manifold of the $SO(3)$ (doubly-connected space) . . . . . . . . . . . . 95
5.7.1 Dynkin Diagrams . . . . . . . . . . . . 110
5.7.2 $\mathfrak{su}(3)$ Weight Diagram for Defining Representation . . . . . . . . . . . . 111
5.7.3 $\mathfrak{su}(3)$ Root Diagram . . . . . . . . . . . . 112
6.4.1 Quark Model . . . . . . . . . . . . 133

# Figures

# Tables

1.1  Klein-4 Group $V$     3
1.2  Symmetric Group $\mathfrak{S}(3)$     3

# Preface

This text grew out of the group theory lectures taught by the first author to undergraduate students at IIT Bombay. While most students attending the course were majoring in physics, there were also students from electrical, mechanical and aerospace engineering streams. The continued interest of students in the topic motivated us to compile the contents into a textbook suitable for a course pitched at the undergraduate level. We also hope that the text will be useful to researchers who wish to familiarize themselves with the basic principles and typical examples of applications of group theory in physics.

In our day-to-day life as well as in the laboratories, we observe patterns with symmetry. These could be in the symmetric shape of the wings of a butterfly, the arrangement of atoms in a molecule, the shape of nuclei as inferred by advanced experimental techniques, among many others. Group theory is a mathematical formulation of such symmetries. While group theory is of interest in mathematics, the prevalence of symmetries in the physical world enable it to find applications in several other disciplines. For instance, the synthesis of molecules in chemistry, crystallographic structures in physics, elasticity properties in continuum mechanics, etc., require the basics of the symmetry principles.

Understanding complex physical systems in nature by solving complicated mathematical equations can be a daunting task. Group theory offers an elegant and powerful framework to extract certain details about such complex systems. For example, the underlying group symmetry can assert whether a given scattering or decay process is allowed or forbidden without performing any explicit computation. This book employs solved examples as well as a variety of exercises (sometimes drawn from sources listed in the Bibliography) to help the reader appreciate the role of group theory in explaining certain experimental observations. We have tried to balance between the formal mathematical aspects and the applications so that the student can effortlessly absorb the group theory arguments behind ad hoc postulates and selection rules. The contents and presentation style enable undergraduate students to connect with the physics they have learned, or will be learning in courses like quantum mechanics, continuum mechanics, atomic and molecular physics, condensed matter physics, nuclear physics, and particle physics in their undergraduate curriculum.

A comprehensive understanding of the concepts in this book will build a strong foundation, preparing the students for further study in diverse areas such as theoretical condensed matter physics or high energy physics, theory including string theory. The first four chapters address discrete group concepts and their applications. The introductory chapter contains definitions and examples, including permutation group. Chapter 2 contains discussion on molecular symmetry, broadly known as point groups. After an initial review of vector spaces and group action on vector spaces, Chapter 3 elaborates on reducible and irreducible representation, character tables, tensor products and their decomposition. This provides the necessary background to determine the vibrational modes of molecules, splitting of degenerate energy level due to impurity or defect which breaks the symmetry of the system partially or completely, and selection rules for transition between energy levels. These applications are elaborated in Chapter 4.

Continuous groups (also known as Lie groups) are discussed in Chapter 5 and Chapter 6. Some of the concepts and notations elaborated in the context of discrete groups are useful while studying systems possessing continuous group symmetry. The key highlights of these two chapters are: (i) Young diagram presentation of irreducible representations for both permutation group as well as special unitary groups which clarifies concepts like tensor products and the decomposition of irreducible representations; (ii) a warm-up on angular momentum algebra which naturally facilitates the description of formal aspects of special unitary group where we introduce weight vectors, root vectors and the Dynkin diagrams; (iii) orthogonal groups and their applications to the energy spectrum of hydrogen atom.

The task of structuring this book in print form would not have been possible without the help of our colleagues and students. We are grateful to Deepak and Himanshu for their timely help in fixing LYX syntax errors which we faced while compiling the chapters. We would like to thank Vivek for helping us with the drawings for some of the figures. We would also like to thank Adiba for helping us in typesetting two sections in LYX package, as well as for proof reading the contents of this book. Our thanks are due also to Himanshu, Aman, Gurbir, Anish, Amihay, Ayaz, Saswati, Zodin, Lata, and Abhishek for their help and valuable comments during various stages in the progress of the book. Besides these students, we greatly appreciate Urjit, Uma, Punit, Pradeep, and Vikram for their comments. Finally, we would like to thank our family members for their encouragement.

We envision that the book will be most suited to teachers offering a one-semester group theory course to undergraduate students. It is also of relevance to researchers embarking on a study of this field. We wish all the readers enjoy their journey through the contents.

# 1

# Introduction

We observe objects in nature to have some pattern or symmetry. In fact, the symmetries inherent in any physical system play a very crucial role in the study of such systems. Group theory is a branch of mathematics that facilitates classification of these symmetries. Hence learning the group theory tools will prove useful to studying applications in physics. Readers will particularly appreciate the power and elegance of the group theoretical techniques in reproducing the experimental observations.

Before delving into its applications, it is important to understand the concept of an abstract group from a purely mathematical standpoint. In this chapter, we present the formal definition of a group and also the notations which will be followed in the rest of the book.

## 1.1 Definition of a Group

**Definition 1.** A group $G$ is a set on which is defined a binary operation called the *product* having the following properties:

1. Closure: For all $a$ and $b$ in $G$, the product $ab$ is in $G$. Here $a$ and $b$ need not be distinct.
2. Associativity: $(ab)c = a(bc)$ for all $a$, $b$ and $c$ in $G$.
3. Existence of Identity: There exists a unique $e$ in $G$ such that $ae = ea = a$ for all $a$ in $G$. $e$ is called the identity element of the group.
4. Existence of Inverse: For every $a$ in $G$ there exists a unique $b$ in $G$ such that $ab = ba = e$. $b$ is called the inverse of $a$ and is conventionally denoted by $a^{-1}$.

In addition to the above mentioned axioms, if it is also true that $ab = ba$ for all $a$ and $b$ in $G$, then $G$ is said to be an *abelian group*. It must be noted that the property of being abelian is special in the sense that not all groups need be abelian. In any group, it is trivial to prove the following statements:

$$\begin{aligned} ax &= bx \Rightarrow a = b \\ xa &= xb \Rightarrow a = b \\ (ab)^{-1} &= b^{-1}a^{-1}. \end{aligned} \qquad (1.1.1)$$

Due to the obvious simplicity of the definition, many familiar sets in mathematics are indeed seen to be examples of groups.

**Example 1.** The set of all integers $Z$ is a group if the group product is taken to be the usual addition of integers. This group is clearly abelian and has an infinite number of elements. □

**Example 2.** The set of all complex numbers $C$ is a group under addition of complex numbers. This group again is abelian and infinite. □

**Example 3.** The set $C - \{0\}$ is an infinite abelian group under the usual multiplication of complex numbers. □

**Example 4.** The set of all $2 \times 2$ matrices with complex entries is an infinite abelian group under matrix addition. □

**Example 5.** The set of all invertible $2 \times 2$ matrices with complex entries is an infinite *non-abelian* group under matrix multiplication. □

A group $G$ that contains a finite number of elements is called a *finite group*. The number of elements in a finite group $G$ is called the *order* of the group and is denoted by $|G|$. For any element $a$ in a group and a positive integer $n$, $a^n$ represents $aa...a$ where there are $n$ factors in the product. Similarly $a^{-n}$ represents $(a^{-1})^n$. Before looking at some examples of finite groups, the following definition may be noted.

**Definition 2.** A subset $S$ of elements of a group $G$ is said to generate $G$ if every element of $G$ can be expressed as a finite product of finite powers of elements (or their inverses) of $S$ in some order. If the set $S$ is finite then the group $G$ is said to be *finitely generated*. The elements of the minimal set $S$ that generates a group $G$ are called the *generators* of the group and $S$ itself is called the *generating set*. A group whose generating set contains a single element is said to be a *cyclic group*.

A group is completely specified if its generators are known along with all the relations that exist between them. It is important to realize that a group may have more than one generating set. In case of finite groups though, the number of elements in all possible generating sets is the same.

**Table 1.1**  Klein-4 Group $V$

| $V$ | $e$ | $a$ | $b$ | $ab$ |
|---|---|---|---|---|
| $e$ | $e$ | $a$ | $b$ | $ab$ |
| $a$ | $a$ | $e$ | $ab$ | $b$ |
| $b$ | $b$ | $ab$ | $e$ | $a$ |
| $ab$ | $ab$ | $b$ | $a$ | $e$ |

**Example 6.** The set $\{e\}$ along with the relation $e^2 = e$ generates the trivial group $\{e\}$ called the identity group of order 1. Henceforth, this group would be represented by the symbol $E$. □

**Example 7.** The set $\{a\}$ along with the relation $a^2 = e$ generates the group $\{e, a\}$ of order 2. □

**Example 8.** The set $\{a\}$ along with the relation $a^n = e$ ($n$ being a positive integer) generates the cyclic group $\{e, a, a^2, \ldots, a^{n-1}\}$ of order $n$. Henceforth, this group would be represented by the symbol $C_n$. □

**Example 9.** The set $\{a, b\}$ along with the relations $a^2 = b^2 = e$ and $ab = ba$ generates the abelian group $V = \{e, a, b, ab\}$. The group $V$ is called the Klein-4 group. It is useful to depict this group in the form of a multiplication table showing all possible products of various group elements as in Table 1.1. Such a table can be drawn for all finite groups. □

**Example 10.** A slightly less trivial example is that of a finite group generated by the set $\{a, b\}$ where $a$ and $b$ satisfy the relations $a^2 = b^3 = e$ and $ab = b^2a$. The generated group $\mathfrak{S}(3) = \{e, a, b, b^2, ab, ab^2\}$ is of order 6 and is non-abelian (Table 1.2). Any product involving a finite number of $a$'s and $b$'s can be reduced to one of the elements of $\mathfrak{S}(3)$ by use of the relations on $a$ and $b$. The notation $\mathfrak{S}(3)$ will be clarified later. □

**Table 1.2**  Symmetric Group $\mathfrak{S}(3)$

| $\mathfrak{S}(3)$ | $e$ | $a$ | $b$ | $b^2$ | $ab$ | $ab^2$ |
|---|---|---|---|---|---|---|
| $e$ | $e$ | $a$ | $b$ | $b^2$ | $ab$ | $ab^2$ |
| $a$ | $a$ | $e$ | $ab$ | $ab^2$ | $b$ | $b^2$ |
| $b$ | $b$ | $ab^2$ | $b^2$ | $e$ | $a$ | $ab$ |
| $b^2$ | $b^2$ | $ab$ | $e$ | $b$ | $ab^2$ | $a$ |
| $ab$ | $ab$ | $b^2$ | $ab^2$ | $a$ | $e$ | $b$ |
| $ab^2$ | $ab^2$ | $b$ | $a$ | $ab$ | $b^2$ | $e$ |

## 1.2 Subgroups

**Definition 3.** A subset $H$ of a group $G$ is called a subgroup if $H$ is a group under the product operation defined on $G$. The identity group $E$ and the group $G$ are both subsets of the group $G$ and are called the *trivial subgroups* of $G$. Other subgroups of $G$ are called *non-trivial subgroups*.

For a subset $H$ of a group $G$ to be a subgroup, $H$ must satisfy all the axioms stated in the Definition 1. If $H$ is closed under the group product and every element of $H$ has its inverse in $H$, then other axioms are automatically satisfied since $H$ is merely a subset of the group $G$. If $G$ is a finite group, then the condition on $H$ to be a subgroup of $G$ is even simpler: $H$ would then just have to be closed under the group product.

Given a finite group $G$, it is easy to find cyclic subgroups of $G$. For example, if $a$ is an element of $G$, all positive integral powers of $a$ are also elements of $G$. $G$ being a finite group implies that there are only finitely many integral powers of $a$ which are distinct elements of $G$. This could happen only if there was some minimum positive integer $n$ for which $a^n = e$, so that further powers of $a$ would merely be repetitions of elements in the set $H = \{e, a, \ldots, a^{n-1}\}$. But then we have generated a cyclic subgroup $H$ of $G$. The order of the subgroup generated by $a$ is also called the order of the element $a$. Evidently, the order of every element of a finite group is finite.

**Example 11.** $\mathfrak{S}(3)$ has $\{e, a\}$, $\{e, ab\}$, $\{e, ab^2\}$, $\{e, b, b^2\}$ as its cyclic subgroups. □

Two subsets of a group $G$ are said to be equal if they contain the same elements. If $A$ and $B$ are two subsets of $G$, then $AB$ is the set of all elements of $G$ which are equal to the product of an element of $A$ with an element of $B$ in that order. It is worth noting that $AB$ and $BA$ need not be equal sets. Suppose that $H$ is a subgroup of $G$. If $a$ is any element of $G$, $Ha$ denotes a subset of $G$ containing elements of the form $ha$ where $h$ runs through all the elements of $H$. Then $Ha$ is called a *left coset* of the subgroup $H$. In the same way, a *right coset* $aH$ can be defined. From hereon, by a coset we mean a left coset. It is evident that $H = He$, and therefore $H$ is one of the cosets of $H$. Every element of $G$ belongs to some coset of $H$ because the coset $Ha$ definitely contains $a$ which could be any arbitrary element of $G$. Thus each and every coset of $H$ contains every possible element of $G$. It is possible that two distinct elements $a$, $b$ of $G$ may belong to the same coset of $H$. This can happen only if there is some $h$ in $H$ such that $ha = b$, or in other words, $ab^{-1}$ is in $H$. Also, two different cosets of $H$ are disjoint. Suppose the two cosets had a common element $a$, then both the cosets must be equal to the coset $Ha$ indicating that the intersection between two different cosets must be a null set. Hence for the finite group $G$, all the cosets of $H$ contain same number of elements and are disjoint whose union is equal to $G$. This proves the important *Lagrange's theorem* which states that the order of every subgroup $H$ of a finite group $G$ divides the order of $G$. Lagrange's theorem does not imply that a subset of a finite group is definitely a subgroup if the number of elements in the subset divides the group order.

# Introduction

A group $G$ may have several subgroups of various orders. Let $H$ be a non-trivial subgroup of $G$. If $a$ is an element of $G$, then $aHa^{-1}$ is a set which contains elements of form $aha^{-1}$ for every $h$ in $H$. It can be verified that $aHa^{-1}$ is also a subgroup of $G$. By choosing a different $a$ each time, we can generate all possible subgroups of the form $aHa^{-1}$. Such subgroups are called *conjugate subgroups*. It is possible that one may get the same subgroup for different choices of $a$. If it turns out that for every choice of $a$, $aHa^{-1} = H$, then $H$ is said to be a *normal subgroup* or *invariant subgroup*. Normal subgroups are special in the sense that they are invariant under conjugation by all the elements of the group, i.e., if $k$ is an element of a normal subgroup $K$, then for all $a$ in $G$, $aka^{-1}$ is again an element of $K$. Further $aKa^{-1} = K \Rightarrow aK = Ka$, i.e., the left and right cosets of a normal subgroup are identical. The trivial subgroups of $G$ are clearly normal. If $G$ is abelian then every subgroup of $G$ is normal. In case $G$ has no non-trivial normal subgroups then $G$ is said to be a *simple group*. Groups of prime order are examples of simple groups.

**Example 12.** Consider the group $\mathfrak{S}(3)$. With the notation $H_a = \{e, a\}$ for the cyclic subgroup generated by $a$ and using the relations on the generators $a$ and $b$, the following can be verified

$$aH_a a^{-1} = H_a;\ (ab)H_a(ab)^{-1} = H_{ab^2};\ (ab^2)H_a(ab^2)^{-1} = H_{ab}.$$

Thus $H_a$, $H_{ab}$, $H_{ab^2}$ are conjugate subgroups. Also, $H_b$ can be seen to be a normal subgroup of $\mathfrak{S}(3)$ and hence $\mathfrak{S}(3)$ is not a simple group. In fact $H_b$ is the only non-trivial normal subgroup of $\mathfrak{S}(3)$. □

The non-trivial normal subgroups of a group play an important role in the study of the group's structure. If $K$ is a normal subgroup of the group $G$, then the set of cosets of $K$ is also a group, called the *factor group* of $G$ with $K$. The factor group is denoted as $G/K$. Let $Ka$ and $Kb$ be two cosets of $K$. Defining the product of the two cosets $(Ka)(Kb)$ to be the set that contains elements of $G$ which are equal to the product of an element of $Ka$ with an element of $Kb$ in that order. Then the elements of $(Ka)(Kb)$ have the form $k_1 a k_2 b = k_1 a k_2 a^{-1} ab = k_1 k_3 ab = k_4\, ab$. When $k_1$ takes all values in $K$, $k_4$ also takes all values in $K$. Thus $(Ka)(Kb) = Kab$ and we have closure in the set of cosets of $K$ under the defined product. Associativity follows from $(KaKb)Kc = KabKc = K(ab)c = Ka(bc) = KaKbc = Ka(KbKc)$. The coset $K$ serves as the identity in $G/K$ and $(Ka)^{-1} = Ka^{-1}$. This proves $G/K$ is a group.

## 1.3 Conjugacy Classes

Suppose $a$ is an element of a group $G$. The set of all elements of $G$ which are equal to $gag^{-1}$ for some choice of $g$ in $G$ is called the *conjugacy class* of the element $a$. The elements of a conjugacy class are said to be conjugate elements of $G$. Conjugate elements, even though distinct, have important common properties. One such property is that if $a$ and $b$ are conjugate then both must have same order. For instance, if $a$ has order $n$ and $b = gag^{-1}$ for some $g$, then

$$b^m = \underbrace{(gag^{-1})(gag^{-1})\ldots(gag^{-1})}_{m \text{ factors}} = ga^m g^{-1}. \tag{1.3.1}$$

The smallest positive integer $m$ for which $ga^m g^{-1}$ is equal to $e$ is clearly $n$. Hence $b$ has the same order as $a$. The above expression also proves that if $a$ and $b$ are conjugate then so are their equal powers. Additionally, if $a$ is conjugate to $b(a = gbg^{-1})$ and $b$ is conjugate to $c(b = hch^{-1})$, then $a = ghch^{-1}g^{-1} = (gh)c(gh)^{-1}$. It follows that $a$ is conjugate to $c$. The property of being conjugate is for this reason *transitive*.

The conjugate elements of $G$ belong to the same conjugacy class. As every element of $G$ is in its own conjugacy class $(a = eae^{-1})$, the whole group $G$ can be divided into several conjugacy classes. It is important that two different conjugacy classes cannot have a common element. For if there was a common element, then that element would be conjugate to all the elements in both the conjugacy classes. By the transitivity property, it follows that the elements in the two classes would be conjugate and therefore must belong to the one and same class. The conjugacy classes of a group decompose the group into mutually exclusive sets. In case of an abelian group, the conjugacy classes consist of the individual group elements.

**Example 13.** Consider $\mathfrak{S}(3)$. Since $geg^{-1} = e$ for all $g$ in $\mathfrak{S}(3)$, $\{e\}$ forms a conjugacy class. Also $bab^{-1} = bab^2 = ab^4 = ab$ and $(b^2)a(b^2)^{-1} = b^2 ab = b^4 a = ba = ab^2$, thus $\{a, ab, ab^2\}$ is a conjugacy class. $aba^{-1} = aba = b^2 \cdot a^2 = b^2$, and it follows $\{b, b^2\}$ is a conjugacy class. We note finally that

$$\mathfrak{S}(3) = \{e\} \cup \{a, ab, ab^2\} \cup \{b, b^2\}.$$

$\square$

## 1.4 Further Examples of Groups

Two important groups are considered here in light of the concepts introduced in previous sections. They are namely: the *Quaternion Group* and the *Dihedral Group*. The quaternion group is a group structure on algebraic entities called *quaternions*. Quaternions were first considered in connection to classical mechanics. They have deep significance as regards mathematical constructs in various physical theories. We do not intend to explore this significance, but we would like to understand what quaternions are in principle. The dihedral group is of importance in the study of symmetries of a regular polygon. We will often encounter the dihedral group in later chapters.

**The Quaternion Group Q**
The quaternion group $Q$ has 8 elements. Explicitly, the group $Q = \{e, s, i, j, k, si, sj, sk\}$. Various elements satisfy the relationships

$$s^2 = e; i^2 = j^2 = k^2 = ijk = s.$$

The symbols $i$, $j$, $k$ can be regarded as imaginary units, $e$ as the real unit and $s$ as negative $e$. The defining relations specify the group completely. The relation $i^2 = ijk$ implies $i = jk$. Likewise, $j = ki$ and $k = ij$. One may note the cyclic nature of these equalities. Furthermore, $i^2 = j^2 \Rightarrow (ji)(ij) = e$. Because $ij = k$ and $(sk)k = e$, it follows $ij = sji$. Likewise, $jk = skj$ and $ki = sik$. The identity element $e$ obviously commutes with all the elements of $Q$, but so does the element $s$. For example, $i = jk \Rightarrow si = sjk$ and $ijk = s \Rightarrow sjk = is$. It follows that $si = is$. In a similar fashion it may be shown $s$ commutes with $j$ and $k$ and hence with all the elements of $Q$. To sum up, we may note the relations that follow from the defining relationships.

$$i = jk, \quad j = ki, \quad k = ij,$$
$$ij = sji, \quad jk = skj, \quad ki = sik,$$
$$si = is, \quad sj = js, \quad sk = ks.$$

The non-trivial subgroups of $Q$ can be of order 2 or 4. The subgroup of order 2 is $\{e, s\}$. The three subgroups of order 4 are $\{e, s, i, si\}$, $\{e, s, j, sj\}$ and $\{e, s, k, sk\}$. Each of the order-4 subgroups is isomorphic to the cyclic group $C_4$. It can be verified that all the subgroups of $Q$ are normal in $Q$. The conjugacy classes of $Q$ can be found from the relationships given above. Since $e$ and $s$ commute with all the elements, they are the only elements in their respective classes. Since $jij^{-1} = sijj^{-1} = si$, it follows $i$ and $si$ are conjugate. Similarly, $j$ and $sj$ are conjugate, and also $k$ and $sk$ are conjugate. The decomposition of $Q$ in conjugacy classes is.

$$Q = \{e\} \cup \{s\} \cup \{i, si\} \cup \{j, sj\} \cup \{k, sk\}.$$

## The Dihedral Group $D_n (n \geq 3)$

The dihedral group $D_n$ has $2n$ elements. It can be generated by the symbols $r$ and $s$ satisfying the relationships as under

$$r^n = s^2 = e, \quad sr = r^{-1}s.$$

Explicitly, $D_n = \{e, s, r, r^2, \ldots r^{n-1}, sr, sr^2, \ldots sr^{n-1}\}$. Any product of a finite number of $r$'s and $s$'s can be reduced to one of the group elements using the defining relationships. Consider

$$r^k s = \underbrace{rr\ldots r}_{k \text{ factors}} s = \underbrace{rr\ldots r}_{k-1 \text{ factors}} sr^{-1} = sr^{-k} = sr^{n-k},$$

and likewise in other cases. The subgroup $\{e, s\}$ is the smallest non-trivial subgroup. The cyclic subgroup generated by $r$, $\{e, r, r^2, \ldots r^{n-1}\}$ is a normal subgroup of $D_n$. If a positive integer $k$ is a divisor of $n$ then $r^k$ will generate a cyclic subgroup of $D_n$. In order to calculate the conjugacy classes, the following relationships are useful:

$$sr^k s = r^{-k},$$
$$r^{-k} sr^k = sr^{2k},$$
$$(sr^t) r^k (sr^t)^{-1} = sr^t r^k r^{-t} s = r^{-k}.$$

In the case when $n$ is $2p+1$, there are $p+2$ classes. The decomposition of $D_{2p+1}$ into classes is

$$D_{2p+1} = \{e\} \cup \underbrace{\{r, r^{2p}\} \cup \{r^2, r^{2p-1}\} \cup \ldots \{r^p, r^{p+1}\}}_{p \text{ classes}} \cup \qquad (1.4.1)$$

$$\cup \{s, sr, sr^2, \ldots sr^{2p}\}.$$

In the case when $n$ is $2p$, there are $p+3$ classes. The decomposition of $D_{2p}$ is seen to be

$$D_{2p} = \{e\} \cup \{r^p\} \cup \underbrace{\{r, r^{2p-1}\} \cup \ldots \{r^{p-1}, r^{p+1}\}}_{p-1 \text{ classes}} \cup \qquad (1.4.2)$$

$$\cup \{s, sr^2, \ldots sr^{2p-2}\} \cup \{sr, sr^3, \ldots sr^{2p-1}\}.$$

## 1.5 Homomorphism of Groups

A *homomorphism* between two groups is a function from one to the other that preserves products. More specifically, a function $\varphi$ from a group $G$ to a group $T$ is a homomorphism if for all $a$, $b$ in $G$

$$\varphi(ab) = \varphi(a)\varphi(b). \qquad (1.5.1)$$

Here the notation $\varphi(a)$ stands for the member of $T$ which is the image of $a$ under the function $\varphi$. If $a$ and $b$ are both chosen to be identity element $e$ of $G$ then $\varphi(e) = [\varphi(e)]^2$ and it follows that the identity of $G$ is mapped to the identity of $T$. Also, if $b = a^{-1}$, then $\varphi(e) = \varphi(a)\varphi(a^{-1})$ and it follows that $\varphi(a^{-1}) = [\varphi(a)]^{-1}$. In other words, under a homomorphism, identity is mapped to identity and inverse is mapped to inverse. The subset $K$ of $G$ which contains all the elements which are mapped to the identity of $T$ is called the *kernel* of the homomorphism $\varphi$. It can be shown that the kernel $K$ is a normal subgroup of $G$. In fact all the normal subgroups of $G$ would cause a homomorphism of $G$ into some group. It can be also shown that all elements of a coset of $K$ are mapped to the same element of $T$.

**Example 14.** Let $C_2 = \{E, A\}$ where $C_2$ is the cyclic group of order 2 in our notation and $E$ here is the identity element of $C_2$. Consider a function $\varphi$ from $\mathfrak{S}(3)$ to $C_2$ defined as

$$\varphi(e) = \varphi(b) = \varphi(b^2) = E,$$

$$\varphi(a) = \varphi(ab) = \varphi(ab^2) = A.$$

Then $\varphi$ is a homomorphism. The kernel of this homomorphism is the normal subgroup $H_b$. $\square$

When the kernel of a homomorphism is simply the trivial group $E$, then the homomorphism is one to one. In this case the two groups are said to be *isomorphic*. Isomorphic groups are essentially the same group and differ merely in the manner of labelling of their elements. In the study of *point groups*, it will be seen that some of them are isomorphic groups. In case of the group $\mathfrak{S}(3)$, the factor group $\mathfrak{S}(3)/H_b$ is isomorphic to the group $C_2$. In notation,

$$C_2 \cong \mathfrak{S}(3)/H_b. \tag{1.5.2}$$

Suppose now that $\varphi_1$ is a homomorphism from $G_0$ into $G_1$ and that $\varphi_2$ is a homomorphism from $G_1$ into $G_2$. It is then possible to compose $\varphi_1$ and $\varphi_2$ together to give a homomorphism $\psi$ from $G_0$ into $G_2$. For $a \in G_0$, define $\psi$ so that

$$\psi(a) = \varphi_2(\varphi_1(a)).$$

It is easy to show that with the above definition, $\psi$ is a homomorphism from $G_0$ into $G_2$.

## 1.6 The Symmetric Group

Consider a set of $n$ distinct letters $\{1, 2, \ldots n\}$ arranged in that order. Any rearrangement of this set of elements is called a *permutation*. For example, if $n = 3$, then all the permutations are

$$\pi_1 = \begin{pmatrix} 1 & 2 & 3 \\ 1 & 2 & 3 \end{pmatrix}; \quad \pi_4 = \begin{pmatrix} 1 & 2 & 3 \\ 3 & 2 & 1 \end{pmatrix}$$

$$\pi_2 = \begin{pmatrix} 1 & 2 & 3 \\ 2 & 1 & 3 \end{pmatrix}; \quad \pi_5 = \begin{pmatrix} 1 & 2 & 3 \\ 2 & 3 & 1 \end{pmatrix}$$

$$\pi_3 = \begin{pmatrix} 1 & 2 & 3 \\ 1 & 3 & 2 \end{pmatrix}; \quad \pi_6 = \begin{pmatrix} 1 & 2 & 3 \\ 3 & 1 & 2 \end{pmatrix}.$$

A permutation that leaves the positions of all the letters unchanged is called the identity permutation. In the example above, $\pi_1$ is the identity permutation. Remember that the exchange amongst columns in the above elements denote the same permutation operation. That is,

$$\pi_5 = \begin{pmatrix} 1 & 2 & 3 \\ 2 & 3 & 1 \end{pmatrix} = \begin{pmatrix} 2 & 1 & 3 \\ 3 & 2 & 1 \end{pmatrix}.$$

A product operation may be defined in the set of permutations so that the product of two permutations is another permutation. The operation of forming the product of two permutations can be most easily understood by considering the concrete case of $n = 3$. Consider the product $\pi_2\pi_5$. $\pi_2$ takes 1 to 2 and $\pi_5$ takes 2 to 3, thus $\pi_2\pi_5$ takes 1 to 3. In this manner the action of $\pi_2\pi_5$ can be ascertained on all the letters

$$\pi_2\pi_5 = \begin{pmatrix} 1 & 2 & 3 \\ 2 & 1 & 3 \end{pmatrix} \begin{pmatrix} 2 & 1 & 3 \\ 3 & 2 & 1 \end{pmatrix} = \begin{pmatrix} 1 & 2 & 3 \\ 3 & 2 & 1 \end{pmatrix} = \pi_4,$$

$$\pi_5\pi_2 = \begin{pmatrix} 1 & 2 & 3 \\ 2 & 3 & 1 \end{pmatrix} \begin{pmatrix} 2 & 3 & 1 \\ 1 & 3 & 2 \end{pmatrix} = \begin{pmatrix} 1 & 2 & 3 \\ 1 & 3 & 2 \end{pmatrix} = \pi_3.$$

Similarly, one may verify that the product so defined is associative. Also $(\pi_2)^{-1} = \pi_2$, $(\pi_3)^{-1} = \pi_3$, $(\pi_4)^{-1} = \pi_4$, $(\pi_5)^{-1} = (\pi_5)^2 = \pi_6$, $(\pi_6)^{-1} = (\pi_6)^2 = \pi_5$. Thus the set of permutations on three letters is a group. In fact, this is the same group as $\mathfrak{S}(3)$ if we identify $\pi_1$ with $e$, $\pi_2$ with $a$, $\pi_5$ with $b$, $\pi_6$ with $b^2$, $\pi_4$ with $ab$ and $\pi_3$ with $ab^2$. Since $\mathfrak{S}(3)$ is the group of all permutations on 3 letters, it is called the *symmetric group* of degree 3. In the general case of permutations on $n$ letters, $\mathfrak{S}(n)$ is called the symmetric group of degree $n$. The order of $\mathfrak{S}(n)$ is $n!$, the total number of permutations of $n$ letters. Subgroups of $\mathfrak{S}(n)$ are called *permutation groups*. The very important *Cayley's Theorem* states that every group is isomorphic to a permutation group which is embedded in some symmetric group. In particular, if $G$ is a group of order $n$, then it is isomorphic to a subgroup of $\mathfrak{S}(n)$. In order to see this, label the elements of $G$ so that $G = \{g_1(=e), g_2, \ldots g_n\}$. Let $g$ be some element of $G$. If every element of $G$ is multiplied with $g$ from right, then we have a permutation $\pi_g$ induced on the letters $\{g_1, g_2, \ldots g_n\}$ which can be written as

$$\pi_g = \begin{pmatrix} g_1 & g_2 & \cdots & g_n \\ g_1g & g_2g & \cdots & g_ng \end{pmatrix}.$$

Another element $h$ of $G$ induces a permutation $\pi_h$ so that the product $\pi_g\pi_h$ is given by

$$\pi_g\pi_h = \begin{pmatrix} g_1 & g_2 & \cdots & g_n \\ g_1g & g_2g & \cdots & g_ng \end{pmatrix} \begin{pmatrix} g_1 & g_2 & \cdots & g_n \\ g_1h & g_2h & \cdots & g_nh \end{pmatrix}$$

$$\Rightarrow \pi_g\pi_h = \begin{pmatrix} g_1 & g_2 & \cdots & g_n \\ g_1gh & g_2gh & \cdots & g_ngh \end{pmatrix}$$

$$\Rightarrow \pi_g\pi_h = \pi_{gh}.$$

This proves that the mapping $g \to \pi_g$ is a homomorphism (Equation 1.5.1). Because two distinct elements of $G$ induce distinct permutations, it follows that the group $G$ and the set of permutations $\{\pi_{g_1}, \pi_{g_2}, \ldots \pi_{g_n}\}$ are isomorphic. This proves the Cayley's Theorem for finite groups.

# Introduction

A permutation can also be represented as a product of disjoint *permutation cycles*. Consider the following permutation on 7 letters.

$$\pi = \begin{pmatrix} 1 & 2 & 3 & 4 & 5 & 6 & 7 \\ 5 & 7 & 3 & 4 & 6 & 1 & 2 \end{pmatrix}.$$

$\pi$ permutes the letters 1, 5, 6 among themselves, and also 2, 7 among themselves while leaving 3 and 4 in their places. This can also be represented as

$$1 \xrightarrow{\pi} 5 \xrightarrow{\pi} 6 \xrightarrow{\pi} 1,\ 2 \xrightarrow{\pi} 7 \xrightarrow{\pi} 2,\ 3 \xrightarrow{\pi} 3,\ 4 \xrightarrow{\pi} 4.$$

Here $(1,5,6)$ is a three-cycle, $(2,7)$ is a two-cycle, $(3)$ and $(4)$ are one-cycles. It is now simpler to write $\pi = (1,5,6)(2,7)(3)(4)$. Since elements in one-cycles are unmoved by the permutation, it is redundant to mention them [for this reason it is conventional to represent the identity permutation simply by the empty cycle $()$]. The decomposition of $\pi$ in disjoint cycles is therefore

$$\pi = (1,5,6)(2,7).$$

The cycle decomposition is particularly useful in carrying out calculations with permutations. Consider another permutation

$$\sigma = \begin{pmatrix} 1 & 2 & 3 & 4 & 5 & 6 & 7 \\ 1 & 4 & 6 & 5 & 7 & 3 & 2 \end{pmatrix} = (3,6)(2,4,5,7).$$

The products $\pi\sigma$ and $\sigma\pi$ can be easily read from the cycle decompositions of the two to be

$$\pi\sigma = (1,7,4,5,3,6);\ \sigma\pi = (1,5,2,4,6,3).$$

It may be noted that $\pi$ (product of a two-cycle and a three-cycle) and $\sigma$ (product of a two-cycle and a four-cycle) have different cycle structure while $\sigma\pi$ and $\pi\sigma$ both has the same six-cycle structure. Further, $(1,5,2,4,6,3), (5,2,4,6,3,1)$ and $(3,1,5,2,4,6)$ represent the same permutations, i.e., positions in a cycle can be changed cyclically without affecting the permutation.

The inverse of a permutation can be obtained by inspection. For example, $\pi$ is a product of a three-cycle and a two-cycle. Since both the cycles are disjoint, $\pi^{-1}$ would be a product of the inverses of the two cycles. A two-cycle like $(2, 7)$ is called a *transposition*. It is obvious that the inverse of a transposition is the transposition itself. As for the three-cycle $(1,5,6)$, its inverse should take 5 to 1, 1 to 6 and 6 to 5, i.e., $(1,5,6)^{-1} = (1,6,5)$. Thus $\pi^{-1} = (1,6,5)(2,7)$, and calculating similarly, $\sigma^{-1} = (3,6)(2,7,5,4)$.

Since a transposition interchanges positions of only two letters, it is the most elementary type of permutation. It is natural to expect that all permutations can be built up from transpositions. It can be checked readily that the $k$-cycle $(1, 2, .., k)$ is equivalent to a product of $k - 1$ transpositions

$$(1, 2, \ldots, k) = (1, 2)(1, 3) \ldots (1, k).$$

As any permutation is a product of disjoint cycles, it can be expressed as a product of transpositions. Depending on whether an odd or an even number of transpositions are involved in the product, the permutation is called *odd* or *even* correspondingly. No permutation is both even and odd.

In the symmetric group $\mathfrak{S}(n)$, it is evident that not all elements would have similar cycle structure. Any permutation in $\mathfrak{S}(n)$ can be written as a product of disjoint $i_k$ $k$-cycles, where $k$ ranges from 1 to $n$ as in

$$\underbrace{\left(a_1^{(1)}\right) \ldots \left(a_{i_1}^{(1)}\right)}_{i_1 1\text{-cycles}} \underbrace{\left(b_1^{(1)}, b_1^{(2)}\right) \ldots \left(b_{i_2}^{(1)}, b_{i_2}^{(2)}\right)}_{i_2 2\text{-cycles}} \ldots$$

$$\ldots \underbrace{\left(t_1^{(1)}, t_1^{(2)}, \ldots t_1^{(n)}\right) \ldots \left(t_{i_n}^{(1)}, t_{i_n}^{(2)}, \ldots t_{i_n}^{(n)}\right)}_{i_n n\text{-cycles}}$$

It is clear that $\Sigma_{k=1}^n k i_k = n$, where some of the $i_k$'s can be 0. Using elementary combinatorics, it can be proved that the total number of such permutations is given by the following expression:

$$n! \prod_{k=1}^n \frac{1}{i_k!(k)^{i_k}}. \tag{1.6.1}$$

The significance of having identical cycle structure is that all such elements are conjugate in $\mathfrak{S}(n)$ and hence belong to the same conjugacy class. Suppose that $\sigma$ and $\pi$ are permutations on same set of letters. Let them be represented by:

$$\sigma = (a_{11} \ldots a_{1s_1})(a_{21} \ldots a_{2s_2}) \ldots (a_{t1} \ldots a_{ts_t}),$$

$$\pi = \begin{pmatrix} a_{11} & \cdots & a_{1s_1} & a_{21} & \cdots & a_{2s_2} & \cdots & a_{t1} & \cdots & a_{ts_t} \\ b_{11} & \cdots & b_{1s_1} & b_{21} & \cdots & b_{2s_2} & \cdots & b_{t1} & \cdots & b_{ts_t} \end{pmatrix}.$$

Consider the action of $\pi^{-1} \sigma \pi$ on $b_{11}$ : $b_{11} \xrightarrow{\pi^{-1}} a_{11} \xrightarrow{\sigma} a_{12} \xrightarrow{\pi} b_{12}$, which means that $\pi^{-1} \sigma \pi$ sends $b_{11}$ to $b_{12}$ and so on. In effect,

$$\pi^{-1} \sigma \pi = (b_{11} \ldots b_{1s_1})(b_{21} \ldots b_{2s_2}) \ldots (b_{t1} \ldots b_{ts_t})$$

and the conclusion is that conjugate elements have the same cycle structure. From the above it is also clear that given two permutations in $\mathfrak{S}(n)$ having identical cycle structures (the RHS of $\sigma$ and $\pi^{-1} \sigma \pi$), one can always find a permutation in $\mathfrak{S}(n)$ (the RHS of $\pi$) so that the first two permutations are related via conjugation by the third.

The conjugacy classes of $\mathfrak{S}(n)$ are distinguished by the cycle structure of elements in the class. Expression 1.6.1 gives the number of permutations in a conjugacy class of a specified cycle structure.

The number of conjugacy classes in $\mathfrak{S}(n)$ is the number of ways in which $n$ can be expressed as a sum of positive integers. For example, if $n = 5$ then there are 7 partitions of $n$, namely $1+1+1+1+1, 1+2+2, 3+2, 1+4, 1+1+1+2, 1+1+3$ and $5$. $\mathfrak{S}(5)$ therefore has 7 conjugacy classes. These conjugacy classes may be conveniently represented by *Young diagrams*. A Young diagram is an arrangement of a certain number of boxes in left justified rows such that the number of boxes in a row is at least as large as the number of boxes in the row below it. Below are listed the Young diagrams for conjugacy classes of $\mathfrak{S}(5)$.

1. The Identity class contains the identity permutation as its only member. The identity is simply the product of 5 one-cycles.

2. Product of 2 two-cycles and 1 one-cycle

3. Product of 1 three-cycle and 1 two-cycle

4. Product of 1 four-cycle and 1 one-cycle

5. Product of 1 two-cycle and 3 one-cycles

6. Product of 1 three-cycle and 2 one-cycles

7. A five-cycle

In the symmetric group $\mathfrak{S}(n)$, the set of all even permutations forms a group called the *alternating group* $\mathfrak{U}(n)$. That $\mathfrak{U}(n)$ is a group follows from the fact that product of two even permutations is an even permutation. The conjugates of an even permutation would have identical cycle structure, and therefore would all be even. It follows that $\mathfrak{U}(n)$ is a normal subgroup of $\mathfrak{S}(n)$. This further implies that the factor group $\dfrac{\mathfrak{S}(n)}{\mathfrak{U}(n)}$ is the set of cosets of $\mathfrak{U}(n)$. Suppose now

$$\frac{\mathfrak{S}(n)}{\mathfrak{U}(n)} = \{\mathfrak{U}(n), \mathfrak{U}(n)(1,2), O_1, \ldots\}.$$

Here $\mathfrak{U}(n)(1,2)$ is a coset of $\mathfrak{U}(n)$ obtained by multiplying all elements of $\mathfrak{U}(n)$ with the transposition (1, 2) and $O_1$ is another coset of $\mathfrak{U}(n)$ containing only odd permutations. Then $O_1 \mathfrak{U}(n)(1,2)$ is also a coset of $\mathfrak{U}(n)$. $O_1 \mathfrak{U}(n)(1,2)$ can contain only even permutations. Because the only coset of $\mathfrak{U}(n)$ containing even permutations is $\mathfrak{U}(n)$ itself, we have $O_1 \mathfrak{U}(n)(1,2) = \mathfrak{U}(n)$. $\mathfrak{U}(n)$ is the identity of the factor group and it follows that $O_1 = [\mathfrak{U}(n)(1,2)]^{-1}$ But $[\mathfrak{U}(n)(1,2)]^{-1} = \mathfrak{U}(n)(1,2)$ and we have the result that there are only two cosets of $\mathfrak{U}(n)$, one being $\mathfrak{U}(n)$ and the other being the set of all odd permutations. Thus the order of the group $\mathfrak{U}(n)$ is $\dfrac{n!}{2}$.

## 1.7 Direct and Semi-direct Products

Two groups $G_1$ and $G_2$ can be used to form a larger group $G$ by a construction called the *direct product* of groups. The direct product of $G_1$ and $G_2$ is denoted as $G = G_1 \times G_2$. Every element of $G$ should be uniquely expressible as a product of an element of $G_1$ and an element of $G_2$. In forming the direct product, $G_1 \times G_2$ is not distinguished from $G_2 \times G_1$. Consequently, every element of $G_1$ should commute with every element of $G_2$. With these properties, it can be seen that $G_1$ and $G_2$ are normal subgroups of $G$. The only element common to $G_1$ and $G_2$ is the identity, for if $g_1$ is in $G_1$ and $g_2$ is in $G_2$, then $g_1 g_2 g_1^{-1} g_2^{-1} = e$ because of commutation. Because $G_1$ is normal $g_1(g_2 g_1^{-1} g_2^{-1})$ is an element of $G_1$, and because $G_2$ is normal $(g_1 g_2 g_1^{-1}) g_2^{-1}$ is an element of $G_2$, and both are equal to the identity element. It is clear that $|G| = |G_1| \times |G_2|$.

**Example 15.** Consider the groups $C_2 = \{e, a\}$ and $C_3 = \{e, b, b^2\}$. From the prescription above, the direct product

$$C_2 \times C_3 = \{e, a, b, b^2, ab, ab^2\}.$$

It may now be noted that the element $ab$ generates the group $C_2 \times C_3$, and order of $ab$ is 6. In fact $C_2 \times C_3$ is isomorphic to $C_6$. In other words

$$C_6 = C_2 \times C_3. \qquad \square$$

In the general case, a direct product may be constructed for any number of groups, the groups themselves being finite or infinite. We do not present this construction here. Finally it may be noted that a finite group $G$ may be expressed as a direct product of its normal subgroups $K_1, K_2, ., K_s$ under the following conditions:

1. $G = K_1 K_2 \ldots K_s$.
2. Every element of $G$ is uniquely expressible as a product of elements from each of the normal subgroups $K_i$.

In a group $G$, let $K$ be a non-trivial normal subgroup. Let $T$ be another non-trivial subgroup of $G$ such that the identity is the only common element in $K$ and $T$. Due to normality of $K$, we have $KT = TK$. Suppose all the group elements $g_i \in G$ can be written as product $k_i t_j$ where $k_i \in K$ and $t_i \in T$. That is, $G = KT$. In this circumstance, $G$ is said to be a *semi-direct product* of $K$ by $T$. The notation for the same is

$$G = K \rtimes T.$$

Notice the positioning of the groups $K$ and $T$ about $\rtimes$. For $t_1$ and $t_2$ in $T$, $Kt_1$ and $Kt_2$ are cosets of $K$. These two cosets will be same only if $t_1 t_2^{-1}$ is in $K$. Because $t_1 t_2^{-1}$ is a member of both $K$ and $T$, from aforementioned conditions on $K$ and $T$ it follows that $t_1 = t_2$. The conclusion is that distinct elements of $T$ are in distinct cosets of $K$ and therefore can be chosen as the coset representatives. Additionally $Kt_i$, with $t_i$ in $T$ exhaust all the cosets of $K$ in $G$ since $G = KT$. If the subgroup $T$ is also normal in $G$ then $G$ is the direct product $K \times T$.

**Example 16.** It was noted in Section 1.6 that the alternating group $\mathfrak{A}(n)$ is a normal subgroup of $\mathfrak{S}(n)$. For $n \geq 3$, let $T$ be the group of order 2 generated by a transposition such as $(1, 2)$. Then

$$\mathfrak{S}(n) = \mathfrak{A}(n) \times T. \qquad \square$$

## Exercises

1. For the following sets $S$ and binary operation $*$ on $S$, determine which of the group axioms stated in Definition 1 are satisfied
   1) $S = N$, $m*n = \max\{m, n\}$ for $m$, $n$ in $N$.
   2) $S = Z$, $m*n = m+n+1$ for $m$, $n$ in $Z$.
   3) $S$ is the set of $n \times n$ invertible real matrices ($n \geq 2$), $A*B = [AB + BA]$.

2. Prove the statements in Equations 1.1.1.

3. Show that a subset $H$ of a finite group $G$ is a subgroup of $G$ if $H$ is closed under the group product.

4. If $H$ is a subgroup of $G$ and $a$ is any arbitrary element of $G$, then show that $aHa^{-1}$ is a subgroup of $G$.

5. Write the multiplication table of $D_3$ and determine the conjugacy classes.

6. If $K$ is a normal subgroup of the group $G$ and $A$ is any subset of elements of $G$, show that $KA = AK$.

7. For subsets $A$ and $B$ of a group $G$, let $AB$ represent the subset of $G$ containing elements $ab$ such that $a$ is in $A$ and $b$ is in $B$. If $A$ and $B$ are subgroups of $G$, and at least one of them is normal in $G$, show that every finite product of elements of $A$ and $B$ is an element of $AB$.

8. With reference to Equation 1.5.1, show that the kernel $K$ of the homomorphism $\phi$ is a subgroup of $G$. Further show that $K$ is a normal subgroup of $G$.

9. In a homomorphism $\phi$, is it always true that the orders of the element $a$ and its image $\phi(a)$ are equal? What if $\phi$ is an isomorphism?

10. Find an isomorphism between a set of positive reals (a group under ordinary multiplication) and all the reals (a group under ordinary addition).

11. Of the groups $C_4$, $D_2$ and the Kelin-4 group $V$, which are the isomorphic groups? Are the groups $D_4$ and $Q$ isomorphic?

12. Consider the group $D_4$. List all its conjugacy classes and its subgroups. Are all the non-trivial subgroups of $D_n$ cyclic?

13. Show that all possible products generated by the Pauli matrices $i\sigma_1$ and $i\sigma_2$ form a group. Write the multiplication table of this group and determine the conjugacy classes. This group is isomorphic to which one of the groups considered in the chapter?

$$\sigma_1 = \begin{pmatrix} 0 & 1 \\ 1 & 0 \end{pmatrix}, \sigma_2 = \begin{pmatrix} 0 & -i \\ i & 0 \end{pmatrix}.$$

14. Show that the composition of homomorphisms defined in Section 1.5 is again a homomorphism.

15. Consider the direct product $C_2 \times C_6$ and state whether it is isomorphic to the cyclic group $C_{12}$.

16. For the symmetric group $\mathfrak{S}(6)$, determine the number of conjugacy classes and the number of permutations belonging to each class. Prepare a Young diagram representation of the classes as illustrated for $\mathfrak{S}(5)$ in the text.

17. Prove that the number of elements belonging to a conjugacy class in $\mathfrak{S}(n)$ is given by Expression 1.6.1.

18. Show that the alternating group $\mathfrak{A}(n)$ is generated by the set of all the three-cycles on $n$ letters.

19. Consider the group $\mathfrak{S}(4)$. Determine its conjugacy classes and the number of elements in each class. Repeat the same exercise for $\mathfrak{A}(4)$. Do the elements appearing in a conjugacy class in $\mathfrak{S}(4)$ appear in the same class in $\mathfrak{A}(4)$?

20. In the group $\mathfrak{S}(4)$, consider the subgroups

$$V = \{(), (1,2), (3,4), (1,2)(3,4)\}$$

and $V_N = \{(), (1,2)(3,4), (1,3)(2,4), (1,4)(2,3)\}$. Show that $V_N$ is normal in $\mathfrak{S}(4)$ while $V$ is not. Realize $\mathfrak{S}(4)$ as a semi-direct product of $V_N$ with another subgroup of $\mathfrak{S}(4)$.

21. Verify $(1, 2, 3, \ldots a, a+1, \ldots n) = (a, a+1, \ldots n)(1,2,3, \ldots a)$.

22. The generators of the symmetric group $\mathfrak{S}(n)$ are $(1,2), (1,2,\ldots n)$. Verify this statement for $\mathfrak{S}(4)$.

23. Let $P_i = (i, i+1)$ denote transposition element of $\mathfrak{S}(n)$ involving neighbouring objects. Show that

$$P_i P_{i+1} P_i = P_{i+1} P_i P_{i+1}; \quad P_i P_j = P_j P_i \text{ for } |i-j| \geq 2.$$

Incidentally, these relations are similar to the defining relations of generators $b_i's$ of *braid group* $\mathfrak{B}(n)$ which emerge as natural symmetry of exchange of neighbouring objects in a two-dimensional plane with $b_i^2 \neq P_i^2 = I$.

# 2

# Molecular Symmetry

In this chapter, we will focus on the symmetries possessed by molecules. By symmetrical transformation of an object, we mean any *rigid* geometric transformation under which the object remains invariant. The three types of rigid transformations are *translations, reflections* and *rotations*. Translations are relevant to the study of symmetries of crystals which is not considered here. For objects of finite size, e.g. a molecule, a translation of the object as a whole definitely does not change its shape and size, but the final position of the object does not exactly overlap its initial position. For this reason, a translation is not an *element of symmetry* of a molecule. However, rotations and reflections or any combinations of the two are elements of symmetry of a molecule. A rotation is a *proper* transformation because it preserves handedness. On the other hand, reflection is *an improper* transformation. We will present symmetry operations and their resemblance to finite group elements discussed in the previous chapter.

## 2.1 Elements of Molecular Symmetry

All the possible symmetry transformations of a molecule can be described in terms of some basic transformations. Of these transformations, the simplest is the *identity transformation E*, which leaves the molecule untouched. This transformation should remind the reader of the identity element of a group.

A molecule may have an *axis of symmetry* about which, if the molecule is rotated by a certain minimum angle (say $\frac{2\pi}{n}$), then it appears just as before the rotation. This transformation is symbolically represented as $C_n$, and the axis is a $n$-fold axis. It is obvious that rotation by the same angle for $n$ consecutive times would bring the molecule to the exact same configuration as it was before the rotations were carried out. The order of the transformation $C_n$ is $n$.

$$C_n^n = E. \tag{2.1.1}$$

Because $C_n^k C_n^{n-k} = E$, $C_n^{n-k}$ may be regarded as the inverse operation of $C_n^k$. It may so happen that a molecule has more than one axis of symmetry. In such a case the axis with maximum $n$ may be called the *principal axis of symmetry*, though this nomenclature is not absolutely necessary. It is conventional to identify the principal axis with the z-axis of the coordinate system. In the special case when there is a two-fold axis of symmetry perpendicular to the principal axis, the two-fold axis would be symbolically represented by $U_2$.

A certain plane through a molecule may exist such that reflection of the molecule in it leaves the molecule indistinguishable from the original. Such a plane is called a *plane of symmetry* A molecule may have more than one such plane. In case the plane contains the principal axis, it is the vertical plane and the transformation corresponding to reflection through this plane is denoted by $\sigma_v$. Reflections in a plane of symmetry perpendicular to the principal axis are denoted by $\sigma_h$ (horizontal). Occasionally, as in case of symmetries of a cube, diagonal planes of symmetry exist, denoted by $\sigma_d$. Order of all the reflection operations is clearly 2.

$$\sigma_v^2 = \sigma_h^2 = \sigma_d^2 = E. \tag{2.1.2}$$

A molecule may have a *roto-reflection symmetry*. This happens if there is a $\sigma_h$ plane perpendicular to a $C_n$ axis. A roto-reflection transformation, $S_n$, can be visualized as a $C_n$ rotation followed by a $\sigma_h$ reflection.

$$S_n = C_n \sigma_h = \sigma_h C_n. \tag{2.1.3}$$

It is obvious from Figure 2.1.1 that the order in which $C_n$ and $\sigma_h$ are carried out is immaterial. If $n$ is an odd integer, then $n$ consecutive roto-reflections on the molecule would merely be equivalent to a reflection in the horizontal plane, for $S_n^n = (C_n \sigma_h)^n = C_n^n \sigma_h^n = \sigma_h$. It follows that $S_n$ is a truly distinct element of symmetry only if $n$ is an even integer, the order of $S_n$ in this case being $n$.

Lastly, a molecule may have *inversion symmetry*, $C_i$. Such a symmetry occurs if the molecule has a *point of symmetry*. The property of such a point is that all the atoms of the molecule lie on lines passing through the point and for every atom lying on some line, there exists another identical atom at the same distance behind the point of symmetry as the first atom is in front of it. Clearly $C_i^2 = E$. Figure 2.1.2 illustrates inversion symmetry.

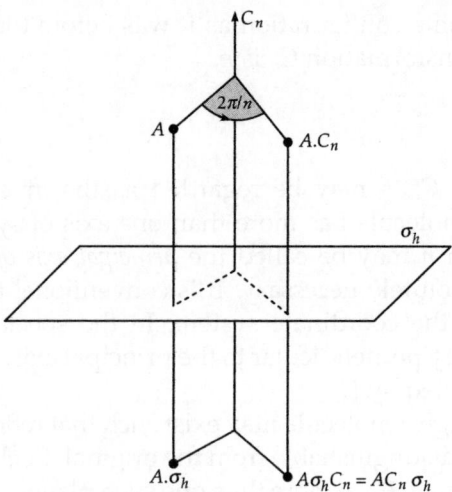

**Figure 2.1.1** Roto-Reflection Symmetry.
The figure shows the action of $C_n\sigma_h$ and $\sigma_h C_n$ on an atom $A$ of some molecule

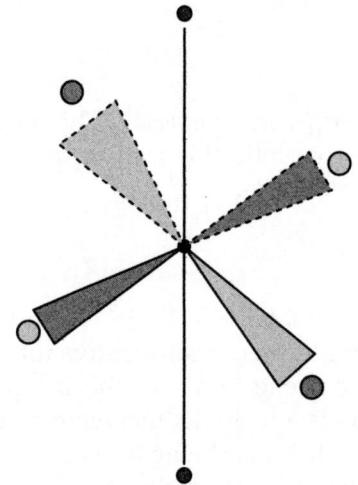

**Figure 2.1.2** Inversion Symmetry.
The central dark sphere is the point of symmetry of this hypothetical molecule

## 2.2 The Symmetry Group of a Molecule

A molecule may possess one or more of the symmetry elements depending upon its structure. Once the symmetry elements of the molecule are identified, one may compose them (in other words, form their products) in various ways and generate all possible symmetry transformations of the molecule. The manner of composition of symmetries is to perform each transformation in the prescribed sequence. For example, if a molecule has a five-fold symmetry axis and a horizontal plane of symmetry, then the operation $C_5^3 \sigma_h$ would mean 3 consecutive rotations of 72° about the principal axis followed by a reflection in the horizontal plane. As each elementary transformation leaves the molecule invariant, it is clear that any transformations composed of elementary transformations are also symmetry transformations. This shows that the set of symmetry transformations is closed under product operation. Equivalently, these transformations imply that all the symmetry axes and symmetry planes of a molecule must coincide at one point.

Note that every elementary symmetry transformation or composed transformation on a molecule permutes all the identical atoms in the molecule (bijection of space points). It is evident that the set of all elementary as well as composed symmetry transformations form a group. The set contains identity element $E$ which corresponds to no symmetry operation. For every elementary symmetry operation $A$, we can show that their inverse operation $A^{-1}$ is in the set. If $C$ and $B$ are elementary symmetry operations, then for the symmetry operation $CB$, the inverse can be defined to be $B^{-1}C^{-1}$. Such a group is an example of a *point group* as there is at least one point in the molecule which remains fixed in all the symmetry transformations.

The number of elements in a point group is finite except for the special case of linear molecules. This follows from the Cayley's Theorem since the point group of a non-linear molecule will be isomorphic to a subgroup of some symmetric group $\mathfrak{S}(n)$. Just like any abstract group, a point group can be decomposed into its conjugacy classes. The conjugacy classes are of importance in the study of *group representations* as will be seen in the next chapter. It is therefore necessary to figure out the conjugacy classes of a given point group. Having identified the various symmetry axes (planes), those axes (planes) are labeled as *equivalent* which can be brought to coincide in direction by a symmetry transformation of the molecule. Symmetry axes of different orders cannot be equivalent. Consider the ammonia molecule shown in Figure 2.2.1. All the $\sigma_v$ planes are equivalent as the alternate planes can be made to coincide with each other after rotation by an angle of $\frac{2\pi}{3}$ about the $C_3$ axis ($\sigma_v \xrightarrow{C_3} \sigma_v'' \xrightarrow{C_3} \sigma_v' \xrightarrow{C_3} \sigma_v$). However, the $\sigma_v$ planes in the water molecule are not equivalent. In the Figure 2.2.2 are shown the symmetry axes of a cube. There are 3 $C_4$ axes, 4 $C_3$ axes and 6 $C_2$ axes. It is easy to verify that all the $C_4$ axes are equivalent, so are all the $C_3$ axes as well as all the $C_2$ axes.

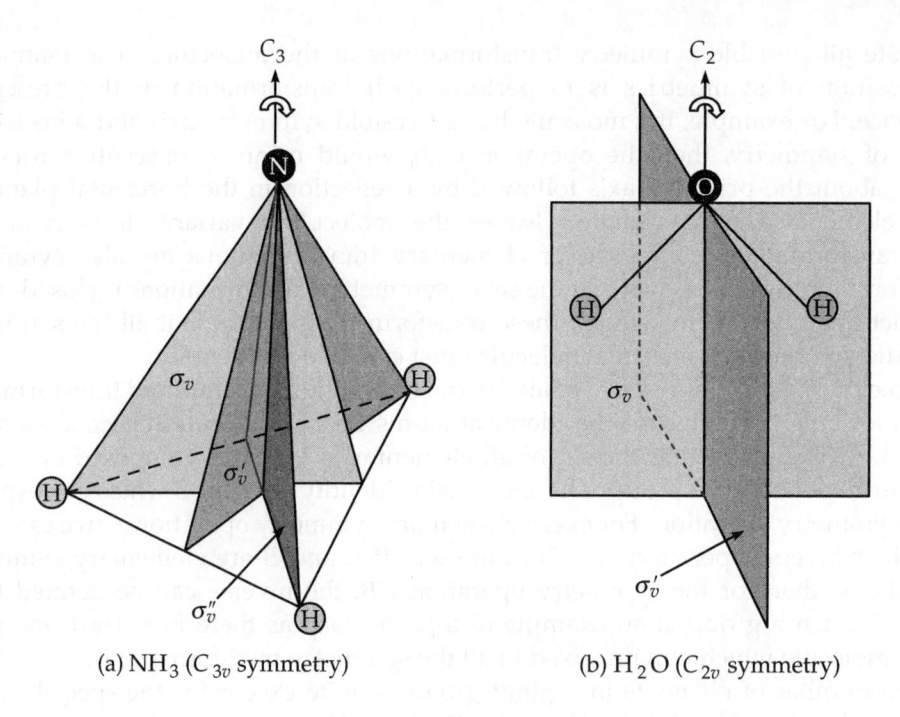

**Figure 2.2.1** Equivalent and Non-equivalent Reflection Planes

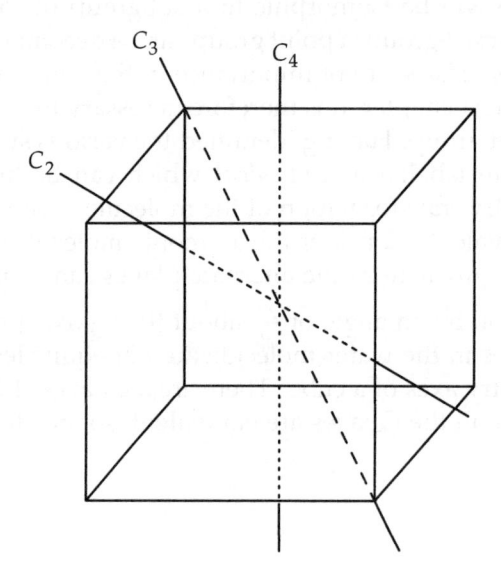

(a) NH$_3$ ($C_{3v}$ symmetry)

(b) H$_2$O ($C_{2v}$ symmetry)

**Figure 2.2.2** Symmetry Axes of a Cube

Equal rotations about equivalent axes are conjugate transformations in the symmetry group of a molecule. In order to see this, suppose $A_1$ and $A_2$ are two equivalent symmetry axes ($C_n$) of a molecule. Then there exists some symmetry transformation $S$ of the molecule which carries $A_1$ to $A_2$. Then the transformation

$SC_n^k S^{-1}$ carries $A_1$ to $A_2$, performs a $C_n^k$ rotation about $A_2$ and bring $A_2$ axis back to $A_1$. This is clearly equivalent to performing a $C_n^k$ rotation about $A_1$ itself. If an axis $C_n$ lies in a symmetry plane $\sigma$, then it is called the *bilateral axis*. For such an axis $C_n^k$ and $C_n^{-k}$ are conjugate. With reference to Figure 2.2.3, the following equalities can be noted:

$$A\sigma = A, \quad A\sigma C_n^{-k} = A',$$
$$\Rightarrow A\sigma C_n^{-k}\sigma^{-1} = A\sigma C_n^{-k}\sigma = A'' = AC_n^k$$
$$\Rightarrow \sigma C_n^{-k}\sigma^{-1} = C_n^k.$$

By a similar construction it is easy to see that an axis is bilateral also when there is a $U_2$ axis (two-fold axis) perpendicular to it (see Exercise 1).

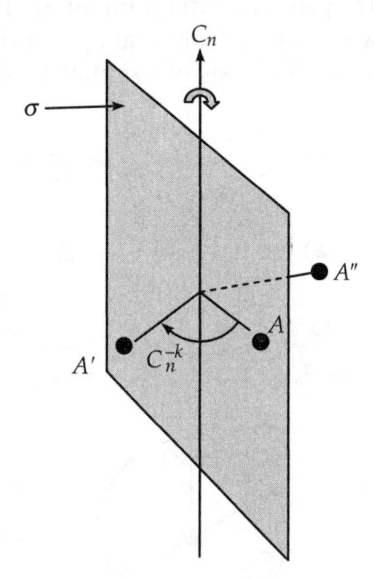

**Figure 2.2.3** Bilateral Axis

## 2.3 Symmetry Point Groups

Finite point groups that describe the symmetries of molecules are considered here. The standard *Schoenflies notation* is used to designate the various groups. The graphical depiction of the groups is done by means of *stereographic projections*. Consider a unit sphere which is divided by the plane of the diagram into two equal parts. In order to form the stereographic projections, we project on the plane of the diagram any general point of the sphere and all its transformed positions obtained by application of the various symmetry transformations of the concerned group. Projections of the points of the upper hemisphere are indicated with a $+$ sign while those of the lower hemisphere are indicated with $\bigcirc$. Further details on stereographic projections are provided as we consider specific groups.

$C_n$ **Group**. A molecule of $C_n$ symmetry has an $n$-fold symmetry axis as its only element of symmetry. This group is clearly cyclic, has $n$ elements and each conjugacy class consists of a single element. The special case of $n = 1$, i.e., $C_1$ corresponds to no symmetry. Figure 2.3.1(a) shows the stereographic projection for a $C_3$ group. Note that the diagram plane cuts the unit sphere along the dotted circle. The central triangle indicates that the axis of symmetry (perpendicular to the diagram plane) is three-fold. A + is taken into the next in the anticlockwise direction upon a $C_3$ operation, and then to the second next upon a $C_3^2$ operation.

$S_{2n}$ **Group**. This is the symmetry group of a molecule having a rotary reflection axis of order $2n$. It is a cyclic group as is evident from the nature of a rotary reflection operation. Thus every element of $S_{2n}$ is the unique member of its conjugacy class. In the Figure 2.3.1(b) is shown the projection diagram for $S_6$. The 6–order rotary reflection axis is indicated by a hollow hexagon. Note that $S_6^2 = C_6\sigma_h C_6\sigma_h = (C_6)^2(\sigma_h)^2 = C_3$, implying the presence of a three-fold axis of symmetry which is indicated by a solid triangle in the diagram.

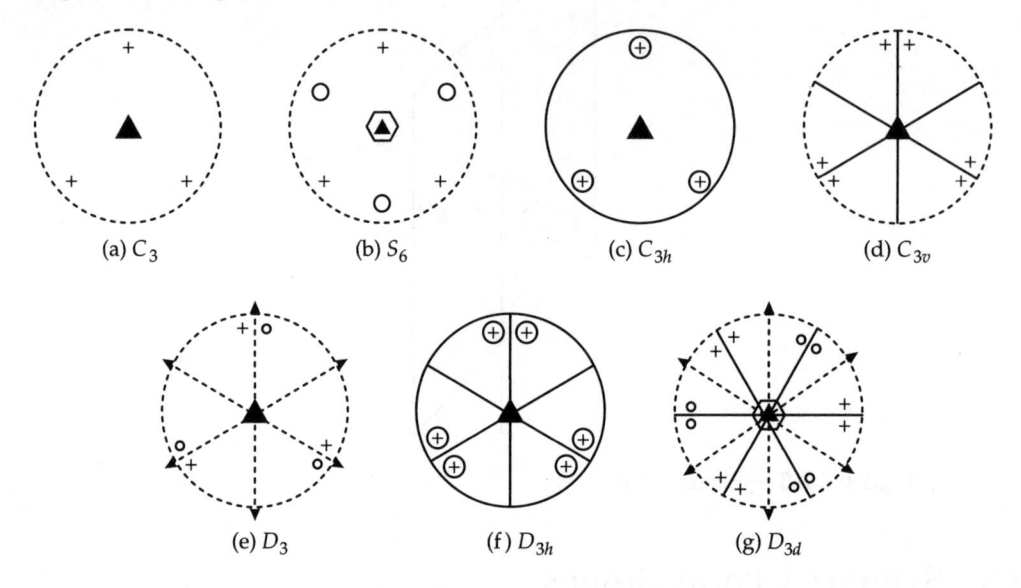

(a) $C_3$  (b) $S_6$  (c) $C_{3h}$  (d) $C_{3v}$

(e) $D_3$  (f) $D_{3h}$  (g) $D_{3d}$

**Figure 2.3.1**  Stereographic Projections

$C_{nh}$ **Group**. If to an $n$-fold axis is added a horizontal plane of symmetry $\sigma_h$, then the $C_{nh}$ group is obtained. This group has $n$ elements of the $C_n$ group and in addition, further $n$ elements of the type $C_n^k \sigma_h$. It is easy to check that the reflection in a horizontal plane and rotation about the principal axis are commuting operations. Therefore $C_{nh}$ is a cyclic group. Consequently every element of $C_{nh}$ is the unique member of its conjugacy class. In the special case of $n = 1$, i.e., $C_{1h}$ contains only the elements $E$ and $\sigma_h$; this group is designated $C_s$. In the stereographic projection for the $C_{3h}$ group, the solid circle indicates a horizontal symmetry plane.

# Molecular Symmetry

The aforementioned groups $C_n$, $S_{2n}$ and $C_{nh}$ are all cyclic groups. For odd $n = 2p + 1$ the groups $S_{2p+1}$ and $C_{(2p+1)h}$ are the same. As abstract groups, it may be noted that two cyclic groups of same order are isomorphic.

$C_{nv}$ **Group.** The $C_{nv}$ group is obtained by adding a $\sigma_v$ plane to a $C_n$ axis. Consequently, a total of $n$ vertical reflection planes appear corresponding to operations $\sigma_v$, $\sigma_v C_n^1$, $\sigma_v C_n^2, \ldots \sigma_v C_n^{n-1}$. The total number of elements in the $C_{nv}$ group is therefore $2n$. The number of conjugacy classes depends upon whether $n$ is odd or even. For $n = 2p$, the vertical planes divide into two sets of equivalent planes. This gives two equivalent classes:

$$\left\{\sigma_v, \sigma_v C_{2p}^2, \ldots \sigma_v C_{2p}^{2p-2}\right\} \ \& \ \left\{\sigma C_{2p}^1, \ldots \sigma_v C_{2p}^{2p-1}\right\},$$

containing $p$ reflections each. The $C_{2p}$ axis is bilateral and thus $C_{2p}^k$ and $C_{2p}^{2p-k}$ are conjugate. This further gives $p - 1$ classes, each containing two rotations and one class containing the rotation $C_{2p}^p = C_2$. Then, there is one class containing the identity $E$. Thus for even $n = 2p$, $C_{nv}$ has $p + 3$ equivalence classes. In a similar manner it may be seen that for odd $n = 2p + 1$, all the reflection planes are equivalent and the number of equivalence classes is $p + 2$. In the stereographic projection for $C_{3v}$, the solid diametrical lines indicate the vertical reflection planes (Figure 2.3.1(d)).

$D_n$ **Group.** The $D_n$ group is obtained by adding a $U_2$ axis to the principal $C_n$ axis. As a result a total of $n$ such $U_2$ axes appear corresponding to operations $U_2$, $U_2 C_n^1, \ldots U_2 C_n^{n-1}$. The total number of elements in $D_n$ is therefore $2n$. The decomposition of the group $D_n$ in conjugacy classes is similar to that of $C_{nv}$ group for both the even and odd $n$. In fact, in may be noted that $D_n$ and $C_{nv}$ are both isomorphic to the abstract dihedral group discussed in Section 1.4. The stereographic projection for $D_3$ shows the $U_2$ axes marked as the double arrowed dotted diametrical lines (Figure 2.3.1(e)).

$D_{nh}$ **Group.** To the group $D_n$, if a plane $\sigma_h$ is added as an element of symmetry, then $n$ vertical planes, $\sigma_v$ appear automatically. Besides the $2n$ elements of group $D_n$, the group $D_{nh}$ contains $n\sigma_v$ reflections as well as $n$ rotary reflections $\sigma_h C_n^k$. Each $\sigma_v$ plane contains the principal axis and a $U_2$ axis. Since $\sigma_h$ commutes with all the other group elements, $D_{nh} = D_n \times C_s$. The number of classes in $D_{nh}$ is twice the number in $D_n$ enumerated as follows: Half of them are same as those of the group $D_n$, while the remainder are obtained by multiplying these by $\sigma_h$. Note that the reflections in $\sigma_v$ planes belong to the same class if $n$ is odd or forms two classes if $n$ is even. Also, the rotary reflection elements $\sigma_h C_n^k$ and $\sigma_h C_n^{n-k}$ are conjugate pairwise.

$D_{nd}$ **Group.** If a diagonal plane $\sigma_d$ bisecting the angle between two adjacent $U_2$ axes of the $D_n$ is added as a symmetry element, then a further $n - 1$ such planes appear. The resultant group is the $D_{nd}$ group containing $4n$ elements. To the $2n$ elements of the $D_n$ group are added $n$ reflections in the diagonal planes and further $n$ transformations of the type $\sigma_d U_2$. The angle between a $U_2$ axis and the adjacent $\sigma_d$ planes is clearly $\pi/2n$.

Figure 2.3.2 illustrates the action of the transformation $\sigma_d U_2$ which is nothing but a rotary reflection. That is, $\sigma_d U_2 = S_{2n}$. All the diagonal reflection planes do contain the $C_n$ axis and hence the principal axis is bilateral. For $n = 2p$, the group $D_{2p,d}$ has $2p + 3$ conjugacy classes: $E$, $C_2$, $(p-1)$ classes of conjugate rotations about principal axis, a single class of $2p U_2$ rotations, a single class of $2p$ reflections $\sigma_d$ and $p$ classes of two rotary reflection transformations $S_{4p}^{2k+1}$ and $S_{4p}^{-2k-1}$ where $k$ can take values 0 to $p-1$. For odd $n = 2p + 1$, inversion is an element of the group and $D_{2p+1,d} = D_{2p+1} \times C_i$.

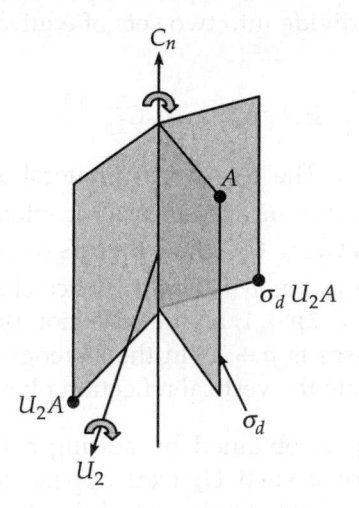

**Figure 2.3.2**  $\sigma_d U_2$ Transformation in $D_{nd}$ Group

**T Group**. The group $T$ consists of all the rotational symmetries of a regular tetrahedron (Figure 2.3.3). A regular tetrahedron is a flat sided solid with 4 triangular faces, all the faces being same. Through each vertex of the tetrahedron, passes a $C_3$ axis. All these axes are equivalent. Additionally there are $C_2$ axes going through midpoints of opposite edges of the tetrahedron. These axes are also equivalent. Hence, the classes of the group $T$ are: $E$, 4 $C_3$ rotations, 4 $C_3^2$ rotations and 3 $C_2$ rotations.

**$T_d$ Group**. This is the full symmetry group of the tetrahedron. Apart from the proper transformations of the group $T$, $T_d$ contains the improper transformations as well. These transformations are obtained by adding 6 diagonal reflection planes. One such plane is shown in Figure 2.3.3. As is evident from the diagram, the plane contains the $C_3$ and $C_2$ axes. Consequently, the $C_3$ axes are bilateral as they contain the mirror planes. Another important feature is that the $C_2$ axes now become $S_4$ axes. This can be seen by rotating the tetrahedron by $\pi/2$ about the $C_2$ axis and reflecting in a perpendicular plane that is midway through the $C_2$ axis. Also $S_4$ axes are bilateral as they are all contained in diagonal planes. Thus the class structure is: $E$, all 8 $C_3$ rotations, all 6 $\sigma_d$ reflections, 3 $C_2(S_4^2)$ rotations, 6 ( $S_4$ and $S_4^3$ are conjugate) rotary reflections.

# Molecular Symmetry

**$T_h$ Group.** To the group $T_d$, if an inversion operation is added, the $T_h$ group is obtained, $T_h = C_i \times T_d$. Consequently the number of conjugacy classes in $T_h$ is twice as many as in $T_d$ (see Exercise 2).

**O Group.** The $O$ group or the octahedral group is the group of all the rotational symmetries of the cube (Figure 2.2.2). The 4 body diagonals are the $C_3$ axes. Then there are 3 $C_4$ axes and 6 $C_2$ axes. Thus the total number of elements of this group is equal to 24 ($4 \times 2 + 3 \times 3 + 6 + 1$). All the $C_4$ axes are equivalent, and so are all the $C_3$ axes and the $C_2$ axes. The $C_4$ axes are bilateral as the $C_2$ axes are perpendicular to them. For the same reason $C_3$ axes are also bilateral. The class structure is: $E$, 6 $C_2$ rotations, 8 rotations of the type $C_3$ and $C_3^2$ 6 rotations of the type $C_4$ and $C_4^3$ and 3 $C_2 (= C_4^2)$ rotations.

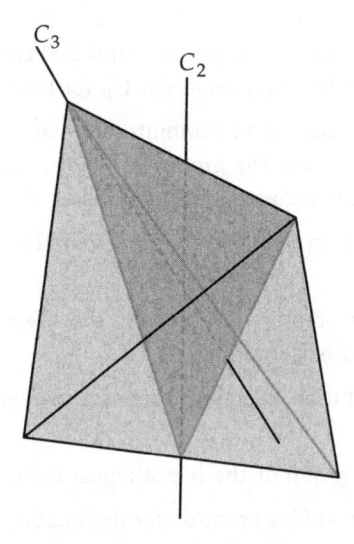

**Figure 2.3.3** Symmetry Axes of the Tetrahedron

**$O_h$ Group.** $O_h$ is the full octahedral group and contains all the possible symmetry transformations of the cube. These include the rotational symmetries contained in the $O$ group as well as those symmetries arising from the fact that the center of the cube is a point of symmetry. Consequently, it can be stated that $O_h = C_i \times O$. $O_h$ has 48 elements and twice as many conjugacy classes as the group $O$.

**Y and $Y_h$ Groups.** These are the symmetries of a regular icosahedron. An icosahedron is a convex polyhedron with twenty triangular faces arranged so that at each vertex of the icosahedron 5 triangular faces meet. The group $Y$ is the group of 60 rotational symmetries of the icosahedron and $Y_h = C_i \times Y$. These symmetries occur rather rarely in molecules.

**Example 17.** Consider the group $C_{3v}$, the symmetry group of the ammonia molecule (Figure 2.2.1). Label the hydrogen atoms by the numbers 1, 2 and 3. Now a $C_3$ rotation sends the atom 1 to atom 2, atom 2 to atom 3 and the atom 3 to atom 1. Thus it is correct to represent the $C_3$ operation by a 3–cycle (1, 2, 3). Now consider the $\sigma_v$ plane that contains the nitrogen atom and the hydrogen atom 3. A reflection in this plane switches the positions of 1 and 2, therefore the operation $\sigma_v$ can be represented by the transposition (1, 2). With this notation, the following equalities are easily verified:

$$C_3 \equiv (1,2,3),\ \sigma_v \equiv (1,2),\ C_3^2 \equiv (1,3,2),$$
$$\sigma_v C_3 \equiv (1,3),\ \sigma_v C_3^2 \equiv (2,3).$$

This way the *permutation representation* of the group $C_{3v}$ is obtained. In fact, it must be noted that the group $C_{3v}$ is isomorphic to the group $\mathfrak{S}(3)$. □

## Exercises

1. Show that an axis of symmetry $C_n$ is bilateral if there exists a $U_2$ axis perpendicular to $C_n$. Find the conjugate rotations about $C_n$ for both cases of $n$ being even and odd.

2. Show that the inversion operation commutes with all other symmetry elements. If $G$ is a point group then show that the group $G \times C_i$ contains twice as many classes as $G$; to each class $A$ of $G$ there correspond two classes $A$ and $C_i A$ in the group $G \times C_i$.

3. Draw the stereographic projection for the group $D_{4h}$. Determine all the conjugacy classes.

4. Find the permutation representation for the Tetrahedral Group $T$. Identify a permutation group that is isomorphic to $T$.

5. Determine the permutation representation for the group $T_d$. Show that $C_{3v}$ is a subgroup of $T_d$.

6. What is the symmetry group of the hypothetical molecule shown in Figure 2.3.4?

7. What could be the Schoenflies notation for designating the symmetry group of the $O_2$ molecule? What could it be for the $CO$ molecule? Are these groups finite?

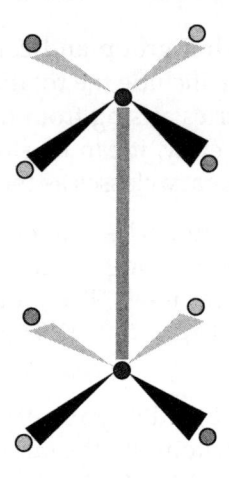

**Figure 2.3.4** Hypothetical Molecule

# 3

# Representations of Finite Groups

In the previous chapter, we observed that group multiplication tables of elements of point groups like $C_{3v}$ and $T$ also apply to the corresponding elements of an appropriate permutation group. In fact, the one to one and onto correspondence between these groups were referred to as isomorphism. Such alternative descriptions of the multiplication table of an abstract group G can also be called as *representations* denoted by $\Gamma(g)$ for every element $g \in G$.

One of the extensively studied approaches towards generating representations of a group G is through the familiar matrices. One associates with each element $g \in G$ a matrix $\Gamma(g)$ such that the group multiplication table is obeyed by the matrices under matrix multiplication. Such representations are particularly useful as they are convenient for performing calculations. We present the necessary aspects of the theory of representations combined with examples. In this chapter our emphasis is on the statements of theorems and the results obtained from their application. Readers interested in detailed proofs are advised to refer to the many available standard texts on group theory. For continuity and clarity, we will now begin with vectors and vector spaces which leads to the construction of matrices acting on such vectors.

## 3.1 Vector Spaces

An abstract *vector space* is an important mathematical structure which is frequently encountered in physics. The elements of a vector space are called *vectors*. The set of vectors forms an abelian group under the addition operation. The identity of this abelian structure is called the *null vector*, denoted usually by 0. Apart from the operation of addition, the vector space is also closed under the operation of *scalar multiplication*. A scalar, for the purposes of this chapter, is simply a complex number. It is natural to assume that for any vector $x$ in a vector space $V$, the complex number 1 times $x$ is same as $x$, i.e., $1x = x$. For scalars $\alpha$, $\beta$ and vectors $x, y$: $(\alpha + \beta)x = \alpha x + \beta x$ as well as $\alpha(x + y) = \alpha x + \alpha y$. Consequently $0x = 0$. The same symbol is used to represent the scalar 0 and the null vector as there is no chance of confusion here. Also, if $x$, $y$ are any two vectors and $\alpha$, $\beta$ are any two complex numbers then the *linear combination* $\alpha x + \beta y$ is also a vector in $V$. A linear combination of any finite number of vectors can be formed in this way. It may be noted that the scalars themselves need not be members of $V$. As described, $V$ is a *complex vector space* since all the scalars belong to the set of complex numbers.

A set of non-null vectors $\{x_i\}_{i=1}^n$ in $V$ is said to be a *linearly independent set* if the only possible linear combination of these vectors that is equal to 0 is the one in which all the multiplicative scalars are themselves 0. The set of all possible linear combinations of the linearly independent vectors $\{x_i\}_{i=1}^n$ is the *linear span* of the linearly independent set. Denote the linear span of $\{x_i\}_{i=1}^n$ by $\mathcal{L}(\{x_i\}_{i=1}^n)$. Clearly $\mathcal{L}(\{x_i\}_{i=1}^n)$ is itself a vector space contained in $V$, and hence is called a subspace of $V$. When $\mathcal{L}(\{x_i\}_{i=1}^n)$ is equal to $V$, then $V$ is said to be of *dimension n* and $\{x_i\}_{i=1}^n$ are said to constitute a *basis* for $V$. Note that $\mathcal{L}(S)$ is defined for any set $S$ of vectors but here the set $\{x_i\}_{i=1}^n$ is assumed to be linearly independent.

Any general vector $x$ in $V$ can be uniquely expressed as a linear combination of the basis vectors. For, if the same vector $x$ could be expressed as $\sum_{i=1}^n \alpha_i x_i$ as well as $\sum_{i=1}^n \beta_i x_i$ where at least one $\alpha_i \neq \beta_i$, then

$$\sum_{i=1}^n (\alpha_i - \beta_i)x_i = 0$$

and it would follow that the basis vectors are not linearly independent. The scalars $\alpha_i$ are called the *components* of the vector $x$ in the basis $\{x_i\}_{i=1}^n$.

From here onwards, it would be assumed that $V$ is of finite dimension, though infinite dimensional vector spaces are quite relevant as well. As the number of vectors in $V$ is infinite (unless of course $V$ contains only the null vector), it may be possible to choose a different set of linearly independent vectors $\{y_j\}_{j=1}^m$ as a basis for $V$. For the concept of dimension of $V$ to be well defined, it is necessary that both the basis sets contain same number of elements, i.e., $m = n$. In order to see this, suppose $\{x_i\}_{i=1}^n$ and $\{y_j\}_{j=1}^m$ are both basis sets and $m > n$. Since $x_1$ can be expressed as a linear combination of $y_j'$s, some $y_k$ can be expressed as a linear combination of $x_1$ and the remaining $y_j'$s. This would mean that the set $B_1 = \{y_1, \ldots, y_{k-1}, x_1, y_{k+1}, \ldots y_m\}$ is a basis set. $x_2$ can be expressed as a linear combination of vectors in $B_1$, and because $x_2$ and $x_1$ are linearly independent, it would follow that some $y_l$ can be expressed as a linear combination of $x_2$ and the remaining elements of $B_1$. Now $y_l$ can be replaced by $x_2$ and a new basis set is obtained to be $B_2 = \{y_1, \ldots, y_{k-1}, x_1, y_{k+1}, \ldots, y_{l-1}, x_2, y_{l+1}, \ldots, y_m\}$. Continuing in this manner, the basis set $B_n$ would contain all the $\{x_i\}_{i=1}^n$ and some $y_j'$s. But because $\{x_i\}_{i=1}^n$ is itself a complete basis for $V$, $B_n$ would no longer be linearly independent which is a contradiction. Thus the conclusion is $m \leq n$. By the same argument, it follows the $n \leq m$ and hence $m = n$.

The subspace $\{0\}$ and the subspace $V$ are the *trivial subspaces* of the vector space $V$. All other subspaces are non-trivial. The dimension of a subspace of $V$ is clearly less than or equal to the dimension of $V$. If there exist subspaces $W_1$ and $W_2$ of $V$ such that every vector $v$ in $V$ can be uniquely expressed as the sum $w_1 + w_2$ where $w_i \in W_i$, then $V$ is said to be a *direct sum* of the subspaces $W_1$ and $W_2$.

$$V = W_1 \oplus W_2. \tag{3.1.1}$$

A *Linear Operator* on a vector space $V$ is a mapping $T$ of $V$ into $V$ such that for all $x$ and $y$ in $V : T(\alpha x + \beta y) = \alpha Tx + \beta Ty$. A consequence of this definition is that $T0 = 0$. Suppose $\{x_i\}_{i=1}^n$ is a basis set for $V$ and $x$ is any element of $V$. By linearity of $T$ it follows that

$$Tx = T\left(\sum_{j=1}^n \alpha_j x_j\right) = \sum_{j=1}^n \alpha_j Tx_j,$$

i.e., the action of $T$ on any vector is determined completely if the action of $T$ is known on the basis set. Suppose now that

$$Tx_j = \sum_{i=1}^n t_{ij} x_i,$$

where $t_{ij}$ are some scalars. Combining the above two

$$Tx = \sum_{i=1}^n \left(\sum_{j=1}^n t_{ij} \alpha_j\right) x_i.$$

Thus the action of $T$ on the vector $x = (\alpha_1, \ldots \alpha_j, \ldots \alpha_n)$ is to transform it into the vector $Tx = (\beta_1, \ldots, \beta_i, \ldots \beta_n)$ whose component $\beta_i = (\sum_{j=1}^n t_{ij} \alpha_j)$. This is most conveniently represented as the matrix product

$$\begin{pmatrix} \beta_1 \\ \vdots \\ \beta_n \end{pmatrix} = \begin{pmatrix} t_{11} & \cdots & t_{1n} \\ \vdots & \ddots & \vdots \\ t_{n1} & \cdots & t_{nn} \end{pmatrix} \begin{pmatrix} \alpha_1 \\ \vdots \\ \alpha_n \end{pmatrix}.$$

The matrix $\{t_{ij}\}$ is identified with the operator $T$. It may be noted that the $j^{th}$ column of the operator matrix contains the components of the vector $Tx_j$. However, the matrix associated with a linear operator is not unique in the sense that the same operator would be represented by a different matrix for a different choice of basis vectors. If $T$ is such that $Tx = Ty \Rightarrow x = y$, then $T$ is a one to one map of $V$ onto $V$. In such a case the matrix of $T$ is an invertible $n \times n$ matrix. It is a fact that the set of all invertible square $n \times n$ matrices form a non-abelian group under matrix multiplication.

It is sometimes necessary to relate the matrix representations of a linear operator $T$ in two different basis sets $\{x_i\}_{i=1}^n$ and $\{y_i\}_{i=1}^n$ of a vector space $V$. Suppose $T_x$ and $T_y$ are the matrices of the operator $T$ in the two bases. The basis vectors $\{x_i\}_{i=1}^n$ themselves must be linearly dependent on the vectors $\{y_i\}_{i=1}^n$ by relations of the form

$$x_i = \sum_{j=1}^n s_{ji} y_j,$$

where $\{s_{ji}\}$ are some scalars. The scalars $s_{ji}$ can be thought of as entries of a matrix $S$. It is clear that $S$ is an invertible matrix, for $\{x_i\}_{i=1}^n$ and $\{y_i\}_{i=1}^n$ form a complete basis. If $x = \sum_{i=1}^n \alpha_i x_i$ is any general vector in $V$, then the representation of $x$ in the basis $\{y_i\}_{i=1}^n$ can now be obtained as

$$x = \sum_{i=1}^n \alpha_i \left( \sum_{j=1}^n s_{ji} y_j \right) = \sum_{j=1}^n \left( \sum_{i=1}^n s_{ji} \alpha_i \right) y_j.$$

It follows that $Sx$ is the representation of $x$ in $\{y_i\}_{i=1}^n$ basis. Because a change of basis should not change the end result of a linear transformation $ST_x x = T_y S x$. In matrix form, this condition can be written as a *similarity transformation*

$$T_x = S^{-1} T_y S. \tag{3.1.2}$$

It may as well be pointed out that the traces (sum of diagonal elements) of matrices that are related by a similarity transformation are equal.

Given an invertible linear operator $T$ on a vector space $V$, the *Eigenvalue Equation* of the operator is

$$Tx = \lambda x.$$

The non-null vectors $x$ satisfying the above equation are called the *eigenvectors* of $T$ and the scalars $\lambda$, the corresponding *eigenvalues*. An important example of an eigenvalue equation is the Schrödinger equation $H\psi = E\psi$ for calculating the stationary states of a closed quantum mechanical system. An operator $T$ which has the same eigenvalue for all $x \in V$ is called a *scaling transformation*.

A vector space $V$ is a *normed space* if for every vector $x$ in $V$ is defined a non-negative real number called the *norm* (or length) of $x$ denoted by $\|x\|$ such that the following properties are satisfied:

1. $\|x + y\| \leq \|x\| + \|y\|$ for all $x$ and $y$ in $V$.
2. $\|\alpha x\| = |\alpha| \|x\|$ for all $x$ in $V$ and all scalars $\alpha$.
3. $\|x\| = 0$ if and only if $x = 0$.

A vector of unit norm is called a *unit vector*.

A vector space $V$ is an *inner product space* (or a *unitary space*) if for all $x$, $y$ in $V$ is defined a complex number $(x, y)$, called their inner product such that the following properties are satisfied:

1. $(x, y) = \overline{(y, x)}$ where the overbar means complex conjugation.
2. $(x, \alpha y + \beta z) = \alpha(x, y) + \beta(x, z)$ for all scalars $\alpha$ and $\beta$.
3. $(x, x) \geq 0$ for all $x$ in $V$.
4. $(x, x) = 0$ if and only if $x = 0$.

Notice that it follows from Property 1 and Property 2 that $(\alpha x, y) = \bar{\alpha}(x, y)$ and also that $(0, x) = (x, 0) = 0$. Two non-null vectors $x$ and $y$ are said to be *orthogonal* if $(x, y) = 0$. Clearly if $(x, y) = 0$ then $(y, x) = 0$. It can be proven that if $x$ and $y$ are orthogonal then they are linearly independent.

A unitary space is naturally a normed space if for a vector $x$, the norm $\|x\| = \sqrt{(x, x)}$. It is easy to check that Property 2 and Property 3 for the norm are satisfied with this definition. In order to verify Property 1, one may note

$$\|x + y\|^2 = (x + y, x + y) = (x, x) + (y, y) + 2\text{Re}\,(x, y)$$
$$\Rightarrow \|x + y\|^2 \leq \|x\|^2 + \|y\|^2 + 2|(x, y)|.$$

It follows from above that $\|x + y\| \leq \|x\| + \|y\|$ will be true if the following, well known *Cauchy–Schwartz inequality* is satisfied:

$$|(x, y)| \leq \|x\|\|y\|. \tag{3.1.3}$$

The Cauchy–Schwartz inequality can be proved as follows. For some scalar $\alpha$ and vectors $x$ and $y$,

$$\|x + \alpha y\|^2 = \|x\|^2 + 2\text{Re}\,[\alpha(x, y)] + |\alpha|^2\|y\|^2 \geq 0.$$

Since the above inequality is true for any arbitrary scalar $\alpha$, it should as well be true in the special case $\alpha = -\frac{(y, x)}{\|y\|^2}$ (assuming $y \neq 0$, otherwise the inequality is trivially true). On substituting this value of $\alpha$, inequality 3.1.3 follows.

The basis set for a unitary space can be chosen in such a way that all the basis vectors are of unit length and are mutually orthogonal. Such a basis is called an *orthonormal basis*. Suppose that $\{x_i\}_{i=1}^n$ is a basis for the unitary space $V$, which is not necessarily orthonormal. From this basis one may construct an orthonormal basis $\{e_i\}_{i=1}^n$ by the *Gram–Schmidt orthogonalisation* process. It can be checked that with the following definitions for $e_i'$s, the basis $\{e_i\}_{i=1}^n$ is indeed orthonormal.

$$e_1 = \frac{x_1}{\|x_1\|}, \; e_2 = \frac{x_2 - (e_1, x_2)e_1}{\|x_2 - (e_1, x_2)e_1\|}, \; e_3 = \frac{x_3 - (e_1, x_3)e_1 - (e_2, x_3)e_2}{\|x_3 - (e_1, x_3)e_1 - (e_2, )e_2\|}, \ldots \quad (3.1.4)$$

$$(e_i, e_j) = \delta_{ij},$$

where $\delta_{ij}$ is the Kronecker delta which assumes the value 1 if $i = j$ and 0 otherwise.

Suppose that $T$ is a linear operator on a unitary space $V$ and $\{e_i\}_{i=1}^n$ an orthonormal basis of $V$. It is then possible to define another operator $T^\dagger$ related to $T$, called the *adjoint* of $T$ by the following relation:

$$t_{ij}^\dagger = \overline{t_{ji}}. \quad (3.1.5)$$

Thus the entries of the adjoint $T^\dagger$ are complex conjugates of the corresponding entries of the transpose of the operator $T$. For this reason the terms *adjoint* and *transpose conjugate* are used interchangeably. The following chains of equalities follow from the definition of the adjoint.

$$(e_i, Te_j) = \left(e_i, \sum_{k=1}^n t_{kj}e_k\right) = \sum_{k=1}^n t_{kj}(e_i, e_k) = \sum_{k=1}^n t_{kj}\delta_{ik} = t_{ij}.$$

$$(T^\dagger e_i, e_j) = \left(\sum_{k=1}^n t_{ki}^\dagger e_k, e_j\right) = \sum_{k=1}^n \overline{t_{ki}^\dagger}(e_k, e_j) = \sum_{k=1}^n \overline{t_{ki}^\dagger}\delta_{kj} = \overline{t_{ji}^\dagger} = t_{ij}.$$

Consequently, $(e_i, Te_j) = (T^\dagger e_i, e_j)$. In general, for any two vectors $x$ and $y$

$$(x, Ty) = (T^\dagger x, y). \quad (3.1.6)$$

An operator is *self-adjoint* if it is equal to its adjoint. In other words $(x, Ty) = (Tx, y)$ for all $x$ and $y$ in $V$. Self-adjoint operators are indeed special as their eigenvalues are real. Let $T$ be self-adjoint and $x$ be an eigenvector with eigenvalue $\lambda$. Then

$$(x, Tx) = (Tx, x)$$

$$\Rightarrow (x, \lambda x) = (\lambda x, x)$$

$$\Rightarrow \lambda(x, x) = \overline{\lambda}(x, x)$$

$$\Rightarrow (\lambda - \overline{\lambda})\|x\|^2 = 0.$$

As $x$ is not a null vector it follows $\lambda$ is real. Self-adjoint operators are particularly important for this reason. In fact, operators corresponding to physical quantities in quantum mechanics are necessarily self-adjoint.

The notion of a *unitary operator* is an important one in physical applications. Such an operator is an invertible linear operator with additional conditions. If $U$ is a unitary operator then its inverse operator is the adjoint of $U$.

$$UU^\dagger = U^\dagger U = 1. \tag{3.1.7}$$

Some of the important properties of unitary operators are worth noting. The norm of a vector remains invariant under a unitary transformation, for $\|Ux\|^2 = (Ux, Ux) = (U^\dagger Ux, x) = (x, x) = \|x\|^2$. If $x$ is an eigenvector of $U$ with eigenvalue $\lambda$, then it follows from the same chain of equalities that $|\lambda|^2 = 1$, i.e., the eigenvalues of a unitary operator are of unit modulus. The rows of a unitary matrix are orthonormal, and so are the columns. The product of two unitary operators (matrix multiplication) is again a unitary operator. Also, the unit matrix is indeed a unitary operator. These properties show that the set of all unitary operators on the vector space $V$ forms a group under matrix multiplication. If $V$ has dimension $n$, then this group is designated as $U(n)$, also referred to as the unitary group of degree $n$. The unitary group and its subgroups will be studied in Chapter 5, *Lie Groups*.

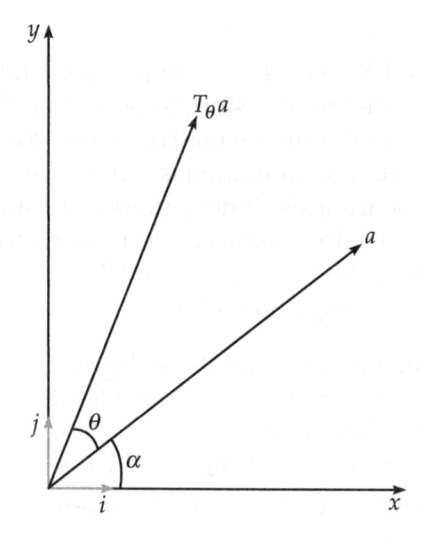

**Figure 3.1.1** $R^2$

The vector space $V$ that is considered here is of the finite dimension. For applications in quantum mechanics, the specialized vector space called *Hilbert space* is sufficient. The Hilbert space is a unitary space with the additional condition of *completeness*. The completeness condition will not be of concern in this text, though it may be implicitly in use. Let it be said that the notions of linear transformations, inverse transformations, eigenvalue equations, adjoints, self-adjoint transformations and unitary operators continue to remain relevant whether the Hilbert space under consideration is finitely dimensional or infinitely dimensional. Also, all the numbered formulae in this section continue to remain valid irrespective.

**Example 18.** Consider the two-dimensional Euclidean plane endowed with a coordinate system consisting of mutually perpendicular *x-y* axes. Then the set of radius vectors of all points in the plane is a two-dimensional *real vector space* $R^2$. A real vector space is one whose multiplicative scalars are chosen from the set of real numbers instead of complex numbers. The reader should be familiar with the *parallelogram law of vector addition* and scaling of vectors via multiplication by real numbers. The unit vectors *i* and *j* (Figure 3.1.1) together form an orthonormal basis.

The linear transformation $R_\theta$ rotates any general radius vector $a$ by an angle of $\theta$ about the origin without changing the length of $a$. If $\|a\|$ is the length of $a$ then in the chosen basis

$$a = \|a\|(i\cos\alpha + j\sin\alpha),$$
$$R_\theta a = \|a\|(i\cos(\alpha+\theta) + j\sin(\alpha+\theta)).$$

On eliminating $\alpha$, the transformation matrix is obtained to be

$$R_\theta = \begin{pmatrix} \cos\theta & -\sin\theta \\ \sin\theta & \cos\theta \end{pmatrix}.$$

It can be verified that indeed $R_\theta(\alpha a + \beta b) = \alpha R_\theta a + \beta R_\theta b$ for any two vectors $a$ and $b$ in $R^2$ and reals $\alpha$ and $\beta$. The inverse transformation of $R_\theta$ is $R_{-\theta}$ which is also equal to the transpose of $R_\theta$ as is evident from the matrix form of $R_\theta$. A transformation whose inverse is the transpose of the transformation is called an *orthogonal transformation*.

$R^2$ is an inner product space if the inner product of $a$ and $b$, $(a, b) = a.b$, where the right hand side is the usual dot product of vectors one learns about in elementary vector algebra. With this definition it follows that $(R_{-\theta}a, b) = (a, R_\theta b)$. Hence $R_{-\theta}$ is the adjoint of $R_\theta$ in the sense of Equation 3.1.6. □

**Example 19.** All homogeneous polynomials of degree two in the variables $x$ and $y$ form a three-dimensional vector space. The basis set for this vector space can be chosen to be $\{x^2, y^2, xy\}$. A general element of this vector space is $\alpha x^2 + \beta y^2 + \gamma xy$ where $\alpha, \beta, \gamma$ are some complex numbers. If the basis set is chosen to be $\{x^2 + y^2, x^2 - y^2, xy\}$, then the same general element can be written as $\frac{\alpha+\beta}{2}(x^2+y^2) + \frac{\alpha-\beta}{2}(x^2-y^2) + \gamma xy$. □

## 3.2 Group Action on Vector Spaces

Given an abstract group $G$ and a vector space $V$, it is often possible to regard the elements of the group as invertible linear operators on the vector space. Once this is done, every element of $G$ can be associated with the matrix of the operator in a fixed basis for $V$. Let $\Gamma$ be a function that maps an element $g$ of $G$ to the matrix $\Gamma(g)$. Then the totality of matrices $\Gamma(g)$ is called a *representation* of $G$ if $\Gamma$ is a homomorphism, i.e.,

$$\Gamma(g_1 g_2) = \Gamma(g_1)\Gamma(g_2). \tag{3.2.1}$$

Notice that the product on the right hand side is a matrix product, while that on the left hand side is the product defined in the group. If the dimension of $V$ is $n$ then the representation $\Gamma$ is said to be of degree $n$. For a given group element $g$, the entries of $\Gamma(g)$ are easily ascertained from the fact that $g$ acts as an invertible linear operator on $V$. Thus the $j^{th}$ column of $\Gamma(g)$ are the coordinates of the transformation of the basis vector $x_j$ of $V$ by the group element $g$ in accordance with the map

$$gx_j \to \sum_{i=1}^{n} [\Gamma(g)]_{ij} x_i.$$

A representation of $\Gamma$ is said to be *faithful* if $\Gamma(g_1) \neq \Gamma(g_2)$ for $g_1 \neq g_2$.

**Example 20.** Consider the point group $C_{3v}$ and the three-dimensional real vector space $\mathbb{R}^3$ spanned by the unit vectors $\{i, j, k\}$ (using the conventional notation). The symmetry axis of $C_{3v}$ points along $k$. The objective is to generate a matrix representation of $C_{3v}$ by letting its elements operate on-$\mathbb{R}^3$. Under the $C_3$ operation (rotation by $\frac{2\pi}{3}$ in the counterclockwise sense about the z-axis), the unit vectors are clearly transformed as follows:

$$i \to -\frac{1}{2}i + \frac{\sqrt{3}}{2}j + 0k,$$

$$j \to -\frac{\sqrt{3}}{2}i - \frac{1}{2}j + 0k,$$

$$k \to 0i + 0j + k.$$

It is therefore plausible to make following associations

$$C_3 \equiv \begin{pmatrix} -\frac{1}{2} & -\frac{\sqrt{3}}{2} & 0 \\ \frac{\sqrt{3}}{2} & -\frac{1}{2} & 0 \\ 0 & 0 & 1 \end{pmatrix}, \quad C_3^2 \equiv \begin{pmatrix} -\frac{1}{2} & -\frac{\sqrt{3}}{2} & 0 \\ -\frac{\sqrt{3}}{2} & -\frac{1}{2} & 0 \\ 0 & 0 & 1 \end{pmatrix}.$$

Now consider the $\sigma_v$ plane to be the z-x plane. Then a $\sigma_v$ reflection leaves $i$ and $k$ invariant and reverses $j$.

$$\sigma_v \equiv \begin{pmatrix} 1 & 0 & 0 \\ 0 & -1 & 0 \\ 0 & 0 & 1 \end{pmatrix}.$$

The matrix forms of $C_3\sigma_v$ and $C_3^2\sigma_v$ can be found from above matrices. The identity element is the $3 \times 3$ identity matrix.

$$E \equiv \begin{pmatrix} 1 & 0 & 0 \\ 0 & 1 & 0 \\ 0 & 0 & 1 \end{pmatrix}, \quad C_3\sigma_v \equiv \begin{pmatrix} -\frac{1}{2} & \frac{\sqrt{3}}{2} & 0 \\ \frac{\sqrt{3}}{2} & \frac{1}{2} & 0 \\ 0 & 0 & 1 \end{pmatrix},$$

$$C_3^2\sigma_v \equiv \begin{pmatrix} -\frac{1}{2} & -\frac{\sqrt{3}}{2} & 0 \\ -\frac{\sqrt{3}}{2} & \frac{1}{2} & 0 \\ 0 & 0 & 1 \end{pmatrix}.$$

$\square$

If $\Gamma(g)$ is the matrix corresponding to group element $g$ in some given basis of $V$, then under a change of basis effected by an invertible matrix $S$, the matrix of $g$ becomes $S^{-1}\Gamma(g)S$ in accordance with Equation 3.1.2.

**Definition 4.** Let $\Gamma$ be a representation of degree $n$ of a group $G$. Then the *character* of an element $g$ of $G$ under the representation $\Gamma$ is the trace of the matrix $\Gamma(g)$.

$$\chi^\Gamma(g) = \text{tr}\Gamma(g) = \sum_{i=1}^{n}[\Gamma(g)]_{ii}. \tag{3.2.2}$$

The symbol $\chi^\Gamma(g)$ would from now onwards stand for the character of the group element $g$ under the representation $\Gamma$.

A consequence of this definition is that if $\Gamma$ and $\Theta$ are two different representations of $G$ related by a similarity transformation, then the character of any element of $G$ has the same value in both the representations. Such representations are *equivalent representations* and are fundamentally not distinguished from each other. A further consequence is that all the group elements which belong to the same conjugacy class have the same character. For if $g_1 = ag_2a^{-1}$, then $\Gamma(g_1) = \Gamma(a)\Gamma(g_2)\Gamma(a)^{-1}$, thereby $\Gamma(g_1)$ and $\Gamma(g_2)$ have the same character. The group character is therefore a *class function*. In the example of the matrix representation of group $C_{3v}$, $\chi(E) = 3$, $\chi(C_3) = \chi(C_3^2) = 0$ and $\chi(\sigma_v) = \chi(C_3\sigma_v) = \chi(C_3^2\sigma_v) = 1$.

## 3.3 Reducible and Irreducible Representations

In a representation $\Gamma$ of a group $G$ over a vector space $V$, it may so happen that some non-trivial subspace $V_1$ of $V$ remains invariant under the group action. By this it is meant that the group elements transform $V_1$ onto $V_1$, however the vectors outside of $V_1$ are transformed to vectors outside $V_1$. When this happens, then $\Gamma$ is also a representation of $G$ on $V_1$ and in the context of $V$, a *subrepresentation* of $G$. With an appropriate choice and ordering of basis vectors, the representing matrix for any general element $g$ of $G$ takes the form

$$\Gamma(g) = \left(\begin{array}{c|c} \Theta(g) & \Pi(g) \\ \hline 0 & \Delta(g) \end{array}\right),$$

where $\Theta(g)$, $\Pi(g)$ and $\Delta(g)$ are fixed size matrix blocks whose entries vary with $g$. The bottom left block has 0 as all entries. A representation such as $\Gamma$ which has a subrepresentation is called a *reducible representation*. A representation with no subrepresentations is called an *irreducible representation*.

A reducible representation is *completely reducible* if it can be decomposed into irreducible representations. If $\Gamma$ were a completely reducible representation then for every $g$

$$\Gamma(g) = \begin{pmatrix} \Gamma^1(g) & 0 & 0 \\ 0 & \ddots & 0 \\ 0 & 0 & \Gamma^k(g) \end{pmatrix}.$$

In other words, the representation matrices of all the group elements can be simultaneously put in a block diagonal form of identical structure, with each block being an irreducible subrepresentation. $\Gamma$ is then a direct sum of the irreducible representations.

$$\Gamma = \bigoplus_{i=1}^{k} \Gamma^i. \tag{3.3.1}$$

It is also obvious from the above decomposition that

$$\chi^\Gamma(g) = \sum_{i=1}^{k} \chi^{\Gamma^i}(g). \tag{3.3.2}$$

It is a theorem that all reducible representations of a finite group obtained via group action on complex or real vector spaces are completely reducible. The interested reader is encouraged to study *Maschke's Theorem* (see Appendix A) in this regard. It may as well be noted that every representation of a finite group over a real or a complex vector space is equivalent to a unitary representation, i.e., the matrices of all the group elements are unitary.

It is fairly obvious from the foregoing that the irreducible representations are the building blocks for all possible representations of a finite group. Therefore a study of the irreducible representations of a group is essential. To begin with, a group has the trivial irreducible representation of degree 1 which is generated when every group element maps the unit vector of a one-dimensional vector space to itself. This representation is also called the *unit representation* and is designated by the symbol $A$. The character of every group element in this representation is equal to 1. It is clear that

$$\sum_{g \in G} \chi^A(g) = |G|, \tag{3.3.3}$$

where summation is over all the elements of the group.

We may recall that the degree of a representation is the dimensionality of the vector space on which the group acts. It may be possible for a group to have non-equivalent irreducible representations of the same degree, and these representations would be treated as distinct. It is a theorem that the degree of an irreducible representation divides the order of the group. Even the number of distinct irreducible representations of a finite group is finite and is equal to the number of conjugacy classes in the group. It has already been noted that group elements in a conjugacy class have the same character under a given representation. Suppose that the finite group $G$ has $r$ conjugacy classes. If $\Gamma_1, \Gamma_2, \ldots \Gamma_r$ are distinct irreducible representations of a group $G$ whose degrees are $\ell_1, \ell_2, \ldots \ell_r$ respectively then

$$\sum_{i=1}^{r} \ell_i^2 = |G|. \tag{3.3.4}$$

The criteria that $G$ has at least one representation of degree 1, that each $\ell_i$ divides $|G|$ and that together the $\ell_i$'s satisfy the above equation put stringent constraints on the possible values of $\ell_i'$s. In most cases, these conditions are sufficient to ascertain the values of $\ell_i'$s.

Let $\Gamma$ and $\Theta$ be two irreducible unitary representations of a finite group $G$. Then the following orthogonality relation holds (read *Schur's Lemma* in this connection which we have briefly discussed in Appendix B) between the entries of the matrices of group elements under the two representations.

$$\sum_{g \in G} [\Gamma(g)]_{ik} \overline{[\Theta(g)]}_{lm} = \frac{|G|}{\ell_\Gamma} \delta_{\Gamma\Theta} \delta_{il} \delta_{km}. \tag{3.3.5}$$

The group characters also follow an orthogonality relation given by

$$\sum_{g \in G} \chi^\Gamma(g) \cdot \overline{\chi^\Theta(g)} = |G| \delta_{\Gamma\Theta}. \tag{3.3.6}$$

As the identity or unit representation $A$ is always an irreducible representation of any group, it follows from Equation 3.3.6 that for any irreducible representation $\Gamma$ other than $A$

$$\sum_{g \in G} \chi^\Gamma(g) = 0. \tag{3.3.7}$$

If $\Gamma = \Theta$ in Equation 3.3.6, then we have

$$\sum_{g \in G} |\chi^\Gamma(g)|^2 = |G|. \tag{3.3.8}$$

Another fact to bear in mind is that the character of the identity element of a group is always equal to the degree of the representation. If $c_1, c_2 \ldots c_r$ represent the classes of $G$ and $n_i$ the number of elements in the conjugacy class $c_i$, then Equation 3.3.6 may once again be written as

$$\sum_{i=1}^{r} \frac{n_i}{|G|} \chi^\Gamma(c_i) \overline{\chi^\Theta(c_i)} = \delta_{\Gamma\Theta}. \tag{3.3.9}$$

As the number of distinct irreducible representations is also equal to number of classes in $G$, one may construct a $r \times r$ matrix whose $k^{th}$ row contains that characters of the classes multiplied by $\sqrt{\frac{n_i}{|G|}}$ in the $k^{th}$ representation. Then the orthogonality of group characters is equivalent to stating that the row vectors of this matrix are orthonormal. Since the columns of such a matrix would also be orthonormal, it immediately follows that

$$\sum_{l=1}^{r} \frac{n_i}{|G|} \chi^{\Gamma_l}(c_i) \overline{\chi^{\Gamma_l}(c_j)} = \delta_{ij}. \tag{3.3.10}$$

## 3.4 Irreducible Representations of Point Groups

In physical applications of the representation theory, it is often only the group characters that are of relevance as against the actual matrices corresponding to the group elements. Equations 3.3.6, 3.3.7 and 3.3.8 prove very useful in the calculation of these characters. The characters of the group classes under various representations are arranged to create a *character table* for the group. A typical character table for representations of a finite group $G$ formally looks like

| $G$ | $n_1 g_1$ | $\ldots$ | $n_r g_r$ | basis |
|---|---|---|---|---|
| $\Gamma_1$ | $\chi^{\Gamma_1}(C_1)$ | $\ldots$ | $\chi^{\Gamma_1}(C_r)$ | |
| $\vdots$ | | $\ddots$ | | |
| $\Gamma_r$ | $\chi^{\Gamma_r}(C_1)$ | | $\chi^{\Gamma_r}(C_r)$ | |

where the symbols have their usual meanings. $n_i g_i$ in the top row states that there are $n_i$ group elements in the conjugacy class $c_i$, and $g_i$ is a representative element of $c_i$. The columns to the right of the character entries contain the basis states of the corresponding representations.

While working with point groups, it is conventional to adopt the *Mulliken symbols* for naming irreducible representations. The principal symmetry axis is taken along the *z-axis*. Representations of degree 1 are identified by symbols A and B, degree 2 representations are identified by the symbol E (not to be confused with the identity element) and degree 3 representations by the symbols F or T. Given a symmetry operation $g$ in the group $G$, the basis vectors of a representation $\Gamma$ get transformed by $\Gamma(g)$. For certain operations, it may happen that the basis vectors are transformed onto themselves, in which case the representation is symmetric. If on the other hand, the basis vectors are reversed, then the representation is antisymmetric. In case of representations $A$, the base functions are symmetric with respect to rotation about principal axis, while they are antisymmetric with respect to rotations about principal axis in case of representations $B$. For representations which are symmetric or antisymmetric with respect to reflections $\sigma_h$, their symbols are marked with one prime or two primes respectively. Representations which are symmetric or antisymmetric with respect to inversions are marked with the letter $g$ (gerade) or $u$ (ungerade) respectively in the subscript to the representation symbol. Following are some examples which illustrate how character tables may be constructed in simple cases.

**Example 21.** Consider the group $C_{2h}$ (Chapter 2, Section 2.3). The number of classes in $C_{2h}$ is 4. Since $1^2 + 1^2 + 1^2 + 1^2 = 4$, all the irreducible representations are of degree 1. The character of identity element of $C_{2h}$ is 1 in all the representations. However, since the order of every other element of $C_{2h}$ is 2, and that each representation is of degree 1, it follows that the character of every element can be either $+1$ or $-1$. The character table follows:

| $C_{2h}$ | $E$ | $C_2$ | $\sigma_h$ | $I$ | | |
|---|---|---|---|---|---|---|
| $A_g$ | 1 | 1 | 1 | 1 | $R_z$ | $x^2; y^2; z^2; xy$ |
| $A_u$ | 1 | 1 | $-1$ | $-1$ | $z$ | |
| $B_g$ | 1 | $-1$ | $-1$ | 1 | $R_x; R_y$ | $xz; yz$ |
| $B_u$ | 1 | $-1$ | 1 | $-1$ | $x; y$ | |

Particular attention should be paid to the names of the irreducible representations. Observe that the character of the inversion operation ($I$) is $+1$ in the gerade representations and $-1$ in the ungerade ones. Similarly the character of $C_2$ operation is $+1$ in A representations and $-1$ in B representations. Two choice of bases have been shown, linear bases in second last column and quadratic bases in the last column. In case of the linear bases, $x$, $y$, $z$ are the polar vectors along the axes and $R_x$, $R_y$, $R_z$ are the axial vectors along the axes. The reflection and inversion properties of these vectors are well known from elementary vector algebra and it is left for the reader to verify that indeed the group operations produce the indicated effects: For example, $C_2 R_x = -R_x$ but $C_2 R_z = R_z$ and so on. In case of quadratic bases it is the signs of the expressions that get affected: For example, $C_2 x^2 = x^2$ but $C_2 xz = -xz$ and so on. Notice that $C_2$ (a rotation by $\pi$ about $z$-axis) changes the sign of $x$ but leaves that of $z$ unchanged. □

**Example 22.** Consider the group $D_{2d}$. This is a group of 8 elements. The 5 conjugacy classes in $D_{2d}$ are: one class containing $E$, one class containing $C_2$ rotation, one containing two rotary reflection transformations $S_4$, one class containing 2 reflections in the diagonal planes $\sigma_d$ and finally one class containing two rotations about the $U_2$ axes. There are therefore 5 irreducible representations of $D_{2d}$. From the equation $1^2 + 1^2 + 1^2 + 1^2 + 2^2 = 8$, it follows that 4 of these representations are of degree 1 and one representation is of degree 2.

| $D_{2d}$ | $E$ | $C_2$ | $2S_4$ | $2U_2$ | $2\sigma_d$ | | |
|---|---|---|---|---|---|---|---|
| $A_1$ | 1 | 1 | 1 | 1 | 1 | | $x^2+y^2, z^2$ |
| $A_2$ | 1 | 1 | 1 | $-1$ | $-1$ | $R_z$ | |
| $B_1$ | 1 | 1 | $-1$ | 1 | $-1$ | | $x^2-y^2$ |
| $B_2$ | 1 | 1 | $-1$ | 1 | 1 | $z$ | $xy$ |
| $E$ | 2 | $-2$ | 0 | 0 | 0 | $(x,y);$ $(R_x, R_y)$ | $xz, yz$ |

In the unit representation $A_1$, all the group characters are 1 as usual. In any of the representations, the group characters satisfy Equations 3.3.7 and 3.3.8. From these it follows that

$$\chi(E) + \chi(C_2) + 2\chi(S_4) + 2\chi(U_2) + 2\chi(\sigma_d) = 0,$$

$$|\chi(E)|^2 + |\chi(C_2)|^2 + 2|\chi(S_4)|^2 + 2|\chi(U_2)|^2 + 2|\chi(\sigma_d)|^2 = 8.$$

The character of identity is 1 in all the other representations of degree 1. Since the orders of $C_2$, $U_2$ and $\sigma_d$ are two, their character values in a one-dimensional representation can be $+1$ or $-1$. The second of the above equations then forces the $|\chi(S_4)| = 1$ and the first of the above equations forces it to be real. The only possible combinations of characters satisfying the above equations for representations of degree 1 are then as shown in the character table. In the representation of degree 2 (i.e., the representation E), the character of identity is clearly 2. The characters of other elements in this representation can now be ascertained easily using orthogonality of the columns of the character table (Equation 3.3.10). For example, the columns of the group elements $E$ and $C_2$ are orthogonal.

$$\overline{\chi^{A_1}(E)}\chi^{A_1}(C_2) + \overline{\chi^{A_2}(E)}\chi^{A_2}(C_2) + \overline{\chi^{B_1}(E)}\chi^{B_1}(C_2)+$$

$$+\overline{\chi^{B_2}(E)}\chi^{B_2}(C_2) + \overline{\chi^{E}(E)}\chi^{E}(C_2) = 0.$$

It follows then that $\chi^E(C_2) = -2$. In the same manner the characters of other elements in the representation E can be easily checked to be 0. Lastly, it may be noted that the principal axis of symmetry in the group is the $S_4$ axis. In accordance with this fact, the symmetric representations $A_1$ and $A_2$ are the ones in which the $\chi(S_4) = 1$, while in the antisymmetric representations $B_1$ and $B_2$, $\chi(S_4) = -1$. □

**Example 23.** Consider the cyclic group $C_3$. Since there are three conjugacy classes in $C_3$, there are 3 irreducible representations of $C_3$. All the representations are of degree 1, as follows from the equation $1^2 + 1^2 + 1^2 = 3$. In the unit representation, all elements have character 1. In a representation of degree 1 (not the unit representation), let the character of rotation $C_3$ be $\omega$.

| $C_3$ | $E$ | $C_3$ | $C_3^2$ | | |
|---|---|---|---|---|---|
| A | 1 | 1 | 1 | $z; R_z$ | $x^2+y^2; z^2$ |
| E $\Big\{$ | 1 | $e^{\frac{2\pi i}{3}}$ | $e^{-\frac{2\pi i}{3}}$ | $(x,y);$ | $(x^2-y^2, xy);$ |
| | 1 | $e^{-\frac{2\pi i}{3}}$ | $e^{\frac{2\pi i}{3}}$ | $(R_x, R_y)$ | $(yz, xz)$ |

Suppose the basis vector for this representation is $\psi$. The action of $C_3$ rotation on $\psi$ is in accordance with $C_3\psi = \omega\psi$. Then

$$C_3^2\psi = C_3(C_3\psi) = C_3(\omega\psi) = \omega C_3\psi = \omega^2\psi.$$

Thus the character of $C_3^2$ in the same representation is $\omega^2$. It follows from Equation 3.3.9 that $1 + \omega + \omega^2 = 0$. Hence $\omega$ is the complex cube root of unity. These representations are shown in the last two rows of the character table. However, instead of identifying them as two separate 1 degree representations, these representations are combined into a single 2 degree representation, identified as E in the first column. The reason for doing so will be made clear when applications are considered. □

**Example 24.** Consider the group $C_{3v}$. The group is non-abelian with 6 members in 3 conjugacy classes. Since $1^2 + 1^2 + 2^2 = 6$, $C_{3v}$ has three irreducible representations, two of which are of degree 1, and one is of degree 2. The calculations in this case are straightforward and the reader can easily verify that indeed the following characters are obtained:

| $C_{3v}$ | $E$ | $2C_3$ | $3\sigma_v$ | | |
|---|---|---|---|---|---|
| $A_1$ | 1 | 1 | 1 | $z$ | $x^2+y^2; z^2$ |
| $A_2$ | 1 | 1 | $-1$ | $R_z$ | |
| E | 2 | $-1$ | 0 | $(x,y); (R_x R_y)$ | $xz; yz$ |

□

## 3.5 The Regular Representation

Consider a finite group $G$ of order $n$ whose elements are $g_1(= E), g_2, \ldots g_n$. Let $\{x_i\}_{i=1}^n$ be a linearly independent set of unit vectors spanning an $n$-dimensional vector space. The group elements can be put in a one to one correspondence with the basis vectors clearly by associating $g_i$ with $x_i$. Then the action of the group element $g$ on this vector space can be formally defined by $x_k = x_i g$ if $g_k = g_i g$. If $g = E$, clearly $x_k = x_i$, i.e., action of the identity element fixes every basis vector. For this reason, $E$ can be associated with the $n \times n$ identity matrix. For any other group element $g$, right multiplication of $G$ with $g$ is a mere permutation of elements of $G$ in which none of the elements of $G$ are fixed. Correspondingly, $g$ transforms any given unit vector in the basis to another distinct unit vector in the basis. It follows that the matrix representation of $g$ would be a $n \times n$ matrix in which each row and each column would contain exactly one non-zero entry which will be equal to 1. This particular representation of G is called the *right regular representation*, which we denote by the symbol $\Gamma^R$. It is clear the degree of $\Gamma^R$ is $n$, $\chi^{\Gamma^R}(E) = n$ and $\chi^{\Gamma^R}(g) = 0$ for $g \neq E$. From orthogonality relations, it follows that $\Gamma^R$ cannot be an irreducible representation.

The immediate task is to decompose $\Gamma^R$ as a direct sum of the irreducible representations of G. In order to do this, assume more generally that $\Gamma$ is some reducible representation of G. Suppose now that

$$\Gamma = \bigoplus_{i=1}^{r} m_i \Gamma^i,$$

where $m_i$ is the multiplicity with which the irreducible representation $\Gamma^i$ appears in $\Gamma$ and $r$ is the number of distinct irreducible representations of G. It follows from above that for any group element $g$

$$\chi^\Gamma(g) = \sum_{i=1}^{r} m_i \chi^{\Gamma^i}(g)$$

$$\Rightarrow \chi^\Gamma(g)\overline{\chi^{\Gamma^k}(g)} = \sum_{i=1}^{r} m_i \chi^{\Gamma^i}(g)\overline{\chi^{\Gamma^k}(g)}$$

$$\Rightarrow \sum_{g \in G} \chi^\Gamma(g)\overline{\chi^{\Gamma^k}(g)} = \sum_{i=1}^{r} m_i \sum_{g \in G} \chi^{\Gamma^i}(g)\overline{\chi^{\Gamma^k}(g)}.$$

Using Equation 3.3.6, it now easily follows that

$$m_k = \frac{1}{|G|} \sum_{g \in G} \chi^\Gamma(g)\overline{\chi^{\Gamma^k}(g)}.$$

The equation above can be used to obtain the multiplicity $m_k$ with which any irreducible representation $\Gamma^k$ appears in a given reducible representation $\Gamma$. In the specific case of $\Gamma = \Gamma^R$, it was earlier noted that $\chi^{\Gamma^R}(E) = n$ and $\chi^{\Gamma^R}(g) = 0$ for $g \neq E$. Since $\chi^{\Gamma^k}(E) = \ell_k$, one has $m_k = \ell_k$ from the above equation. Each irreducible representation of G appears as many times as its degree in the right regular representation of G. Additionally, since $\chi^{\Gamma^R}(E) = \sum_{i=1}^{r} m_i \chi^{\Gamma^i}(E)$, Equation 3.3.4 follows as a consequence. Finally it may be noted that the trivial representation (of degree 1) occurs once in $\Gamma^R$.

## 3.6 Tensor Product of Representations

It is possible to generate reducible representations of a group from its irreducible ones. This can be done by forming what is called the *tensor product* of the irreducible representations. Let $G$ be a finite group which has $\Gamma_1$ and $\Gamma_2$ as irreducible representations. $\Gamma_1$ is a representation of $G$ on a unitary space $V$ spanned by the basis $\{x_i\}_{i=1}^{n_1}$ whereas $\Gamma_2$ is a representation of $G$ on a unitary space $W$ spanned by the basis $\{y_i\}_{i=1}^{n_2}$. The tensor product of $V$ and $W$, denoted by $V \otimes W$, is another vector space spanned by the basis $\{x_i \otimes y_j\}_{i=1,j=1}^{n_1,n_2}$. The dimension of $V \otimes W$ is clearly equal to the product of the dimensions of $V$ and $W$. Once the product space has been defined, it remains to construct a representation $\Gamma$ of $G$ on $V \otimes W$. The representation $\Gamma$ is called the tensor product of the representations $\Gamma_1$ and $\Gamma_2$ and symbolically $\Gamma = \Gamma_1 \otimes \Gamma_2$. For completing the definition, $\Gamma(g)$ should be simply related to $\Gamma_1(g)$ and $\Gamma_2(g)$ for all $g$ in $G$. This essentially requires the specification of action of $g$ on every basis vector of $V \otimes W$. For the basis vector $x_i \otimes y_j$ in $V \otimes W$ let

$$g(x_i \otimes y_j) \to g(x_i) \otimes g(y_j)$$

$$g(x_i \otimes y_j) \to \left(\sum_{k=1}^{n_1}[\Gamma_1(g)]_{ki}x_k\right) \otimes \left(\sum_{l=1}^{n_2}[\Gamma_2(g)]_{lj}y_l\right)$$

$$g(x_i \otimes y_j) \to \sum_{k=1}^{n_1}\sum_{l=1}^{n_2}[\Gamma_1(g)]_{ki}[\Gamma_2(g)]_{lj}x_k \otimes y_l. \tag{3.6.1}$$

The above mapping defines the action of $g$ on a basis state $x_i \otimes y_j$ of $V \otimes W$. It remains to be shown that with such a definition, $\Gamma$ is indeed a representation of $G$ on $V \otimes W$, or in other words, Equation 3.2.1 is satisfied. The rows and columns of the matrix $\Gamma(g)$ can be indexed by the basis states $x_i \otimes y_j$. Equation 3.2.1 is true if

$$[\Gamma(g_1 g_2)]_{x_u \otimes y_v, x_s \otimes y_i} = [\Gamma(g_1)\Gamma(g_2)]_{x_u \otimes y_v, x_s \otimes y_i}.$$

The verification of the above equality is trivial after noticing

$$[\Gamma(g)]_{x_u \otimes y_v, x_s \otimes y_i} = [\Gamma_1(g)]_{us}[\Gamma_2(g)]_{vi}, \tag{3.6.2}$$

which follows from the mapping defined in 3.6.1. The character of $g$ can be easily obtained by summing the diagonal elements. A direct consequence of the above is that

$$\chi^{\Gamma=\Gamma_1 \otimes \Gamma_2}(g) = \chi^{\Gamma_1}(g)\chi^{\Gamma_2}(g). \tag{3.6.3}$$

## 3.7 Decomposition of Reducible Representations

A reducible representation $\Gamma$ of a group $G$ over a unitary space $V$ can be expressed as a direct sum of the irreducible representations $\Gamma^\alpha$ of $G$ in which each irreducible representation appears with a certain multiplicity $m_\alpha$.

$$\Gamma = \bigoplus_{\alpha=1}^{r} m_\alpha \Gamma^\alpha. \tag{3.7.1}$$

In Section 3.5, the formula for $m_\alpha$ was obtained which is restated below for emphasis.

$$m_\alpha = \frac{1}{|G|} \sum_{g \in G} \chi^\Gamma(g) \overline{\chi^{\Gamma^\alpha}(g)}. \tag{3.7.2}$$

It may be desirable sometimes to not only know the multiplicities of the various irreducible representations in $\Gamma$ but also the subspace of $V$ which is left invariant by the irreducible representation $\Gamma^\alpha$ (i.e., the subspace of $V$ on which $\Gamma^\alpha$ is an irreducible representation of $G$).

Let $V$ be a unitary space of which $U$ and $W$ are proper subspaces such that $V = U \oplus W$. Then any vector $v$ in $V$ can be expressed uniquely as a sum of vectors $u$ and $w$ which belong to $U$ and $W$ respectively. The projection operator $P_U$ is a linear operator on $V$ with values in the subspace $U$ such that

$$P_U v = u.$$

Similarly $P_W$ may be defined so that $P_W v = w$. Since the projection operator is a mapping onto a proper subspace, it is not invertible. The projection operator essentially projects a vector in $V$ on a subspace (for which the operator is defined). As a trivial example, the projection operators that project out the $x$, $y$ and $z$ components of a vector in $\mathbb{R}^3$ are

$$P_x = \begin{pmatrix} 1 & 0 & 0 \\ 0 & 0 & 0 \\ 0 & 0 & 0 \end{pmatrix}; \quad P_y = \begin{pmatrix} 0 & 0 & 0 \\ 0 & 1 & 0 \\ 0 & 0 & 0 \end{pmatrix}; \quad P_z = \begin{pmatrix} 0 & 0 & 0 \\ 0 & 0 & 0 \\ 0 & 0 & 1 \end{pmatrix}.$$

The problem of finding the subspace of $V$ which is left invariant by the irreducible representation $\Gamma^\alpha$ (Equation 3.7.1) reduces to constructing the project operator $P_{\Gamma^\alpha}$. In the orthogonality relation (Equation 3.3.5)

$$\sum_{g \in G} \overline{[\Gamma^\alpha(g)]}_{ik} [\Gamma^\beta(g)]_{lm} = \frac{|G|}{\ell_{\Gamma^\alpha}} \delta_{\Gamma^\alpha \Gamma^\beta} \delta_{il} \delta_{km},$$

if $i$ is set equal to $k$ and summation over all $i$ is performed, then the relation reduces to

$$\frac{\ell_{\Gamma^\alpha}}{|G|} \sum_{g \in G} \overline{\chi^\alpha(g)} [\Gamma^\beta(g)]_{lm} = \delta_{\Gamma^\alpha \Gamma^\beta} \delta_{lm}.$$

It is evident that when $\Gamma^\alpha$ and $\Gamma^\beta$ are the same representations then the right hand side of the above is the identity matrix of order $\ell_\alpha \times \ell_\alpha$ and in other cases the right hand side is the null matrix. This important property motivates the following definition of the projection operator $P_{\Gamma^\alpha}$ on the reducible representation $\Gamma$.

$$P_{\Gamma^\alpha} = \frac{\ell_{\Gamma^\alpha}}{|G|} \sum_{g \in G} \overline{\chi^{\Gamma^\alpha}(g)} \Gamma(g). \tag{3.7.3}$$

In the direct sum in Equation 3.7.1, upon action of $P_{\Gamma^\alpha}$ only those blocks in the matrix $\Gamma(g)$ will be non-zero which correspond to subspaces on which $\Gamma^\alpha$ gives a representation of $G$.

**Example 25.** Consider the group $C_{3v}$. In an earlier example (Section 3.2, Example 20), the explicit form of the matrices for various group elements was calculated by considering the group's action on $\mathbb{R}^3$. The $k$ vector was left invariant under all transformations and in fact, the representation generated was of degree 2. For reference, the representing matrices of various group elements of $C_{3v}$ in the irreducible representation $E$ are given below:

$$E(E) = \begin{pmatrix} 1 & 0 \\ 0 & 1 \end{pmatrix}; \qquad E(C_3) = \begin{pmatrix} -\frac{1}{2} & -\frac{\sqrt{3}}{2} \\ \frac{\sqrt{3}}{2} & -\frac{1}{2} \end{pmatrix};$$

$$E\left(C_3^2\right) = \begin{pmatrix} -\frac{1}{2} & \frac{\sqrt{3}}{2} \\ -\frac{\sqrt{3}}{2} & -\frac{1}{2} \end{pmatrix}; \qquad E(\sigma_v) = \begin{pmatrix} 1 & 0 \\ 0 & -1 \end{pmatrix};$$

$$E(C_3\sigma_v) = \begin{pmatrix} -\frac{1}{2} & \frac{\sqrt{3}}{2} \\ \frac{\sqrt{3}}{2} & \frac{1}{2} \end{pmatrix}; \qquad E\left(C_3^2\sigma_v\right) = \begin{pmatrix} -\frac{1}{2} & -\frac{\sqrt{3}}{2} \\ -\frac{\sqrt{3}}{2} & \frac{1}{2} \end{pmatrix}.$$

Let $\Gamma = E \otimes E$ be the tensor product of the irreducible representation of degree 2 with itself. From Equation 3.6.3, it follows that $\chi^\Gamma(E) = 4$, $\chi^\Gamma(C_3) = \chi^\Gamma(C_3^2) = 1$ and $\chi^\Gamma(\sigma_v) = \chi^\Gamma(\sigma_v C_3) = \chi^\Gamma(\sigma_v C_3^2) = 0$. It is now possible to decompose $\Gamma$. The multiplicity of various irreducible representations in $\Gamma$ can be calculated by application of Equation 3.7.2.

Multiplicity of $A_1 = \frac{1}{6}[1 \times 4 \times 1 + 2 \times 1 \times 1 + 3 \times 0 \times 1] = 1$.

Multiplicity of $A_2 = \frac{1}{6}[1 \times 4 \times 1 + 2 \times 1 \times 1 + 3 \times 0 \times 0] = 1$.

Multiplicity of $E = \frac{1}{6}[1 \times 4 \times 2 + 2 \times 1 \times -1 + 3 \times 0 \times 0] = 1$.

It may be noted that $\Gamma = A_1 \oplus A_2 \oplus E$. Each of the irreducible representations of $C_{3v}$ occur only once in $\Gamma$. From the matrices of the representation E, the matrices of the representation $\Gamma$ can be calculated by using Equation 3.6.2. The basis for the tensor product space in this case can be taken as $i_1 \otimes i_2$, $i_1 \otimes j_2$, $j_1 \otimes i_2$ and $j_1 \otimes j_2$ where

# Representations of Finite Groups

$\{i_1, j_1\}$ is the basis of one $\Gamma^E$ and $\{i_2, j_2\}$ is the basis for the other. The rows and columns of the matrices in $\Gamma$ representation are indexed in the order of the basis states of the tensor product space as mentioned in the previous line. With this convention, following matrices in $\Gamma$ representation are obtained:

$$\Gamma(E) = \begin{pmatrix} 1 & 0 & 0 & 0 \\ 0 & 1 & 0 & 0 \\ 0 & 0 & 1 & 0 \\ 0 & 0 & 0 & 1 \end{pmatrix},$$

$$\Gamma(C_3) = \begin{pmatrix} \frac{1}{4} & \frac{\sqrt{3}}{4} & \frac{\sqrt{3}}{4} & \frac{3}{4} \\ -\frac{\sqrt{3}}{4} & \frac{1}{4} & -\frac{3}{4} & \frac{\sqrt{3}}{4} \\ -\frac{\sqrt{3}}{4} & -\frac{3}{4} & \frac{1}{4} & \frac{\sqrt{3}}{4} \\ \frac{3}{4} & -\frac{\sqrt{3}}{4} & -\frac{\sqrt{3}}{4} & \frac{1}{4} \end{pmatrix},$$

$$\Gamma(C_3^2) = \begin{pmatrix} \frac{1}{4} & -\frac{\sqrt{3}}{4} & -\frac{\sqrt{3}}{4} & \frac{3}{4} \\ \frac{\sqrt{3}}{4} & \frac{1}{4} & -\frac{3}{4} & -\frac{\sqrt{3}}{4} \\ \frac{\sqrt{3}}{4} & -\frac{3}{4} & \frac{1}{4} & -\frac{\sqrt{3}}{4} \\ \frac{3}{4} & \frac{\sqrt{3}}{4} & \frac{\sqrt{3}}{4} & -\frac{1}{4} \end{pmatrix},$$

$$\Gamma(\sigma_v) = \begin{pmatrix} 1 & 0 & 0 & 0 \\ 0 & -1 & 0 & 0 \\ 0 & 0 & -1 & 0 \\ 0 & 0 & 0 & 1 \end{pmatrix},$$

$$\Gamma(C_3\sigma_v) = \begin{pmatrix} \frac{1}{4} & -\frac{\sqrt{3}}{4} & -\frac{\sqrt{3}}{4} & \frac{3}{4} \\ -\frac{\sqrt{3}}{4} & -\frac{1}{4} & \frac{3}{4} & \frac{\sqrt{3}}{4} \\ -\frac{\sqrt{3}}{4} & \frac{3}{4} & -\frac{1}{4} & \frac{\sqrt{3}}{4} \\ \frac{3}{4} & \frac{\sqrt{3}}{4} & \frac{\sqrt{3}}{4} & \frac{1}{4} \end{pmatrix},$$

$$\Gamma(C_3^2\sigma_v) = \begin{pmatrix} \frac{1}{4} & \frac{\sqrt{3}}{4} & \frac{\sqrt{3}}{4} & \frac{3}{4} \\ \frac{\sqrt{3}}{4} & -\frac{1}{4} & \frac{3}{4} & -\frac{\sqrt{3}}{4} \\ \frac{\sqrt{3}}{4} & \frac{3}{4} & -\frac{1}{4} & -\frac{\sqrt{3}}{4} \\ \frac{3}{4} & -\frac{\sqrt{3}}{4} & -\frac{\sqrt{3}}{4} & \frac{1}{4} \end{pmatrix}.$$

Finally the projection operators for the irreducible representations in $\Gamma$ may be calculated from these matrices, the characters of irreducible representations and Equation 3.7.3. For example, all the characters in $A_1$ are unit, and Equation 3.7.3 takes the form

$$P_{A_1} = \frac{1}{6}\left[\Gamma(E) + \Gamma(C_3) + \Gamma(C_3^2) + \Gamma(\sigma_v) + \Gamma(C_3\sigma_v) + \Gamma(C_3^2\sigma_v)\right].$$

In a similar fashion, the projection operators for other irreducible representations may be obtained. The results follow:

$$P_{A_1} = \begin{pmatrix} \frac{1}{2} & 0 & 0 & \frac{1}{2} \\ 0 & 0 & 0 & 0 \\ 0 & 0 & 0 & 0 \\ \frac{1}{2} & 0 & 0 & \frac{1}{2} \end{pmatrix}, \quad P_E = \begin{pmatrix} \frac{1}{2} & 0 & 0 & -\frac{1}{2} \\ 0 & \frac{1}{2} & \frac{1}{2} & 0 \\ 0 & \frac{1}{2} & \frac{1}{2} & 0 \\ -\frac{1}{2} & 0 & 0 & \frac{1}{2} \end{pmatrix},$$

$$P_{A_2} = \begin{pmatrix} 0 & 0 & 0 & 0 \\ 0 & \frac{1}{2} & -\frac{1}{2} & 0 \\ 0 & -\frac{1}{2} & \frac{1}{2} & 0 \\ 0 & 0 & 0 & 0 \end{pmatrix}.$$

By letting these projection operators act on any general vector in the tensor product space spanned by $i_1 \otimes i_2$, $i_1 \otimes j_2$, $j_1 \otimes i_2$ and $j_1 \otimes j_2$, the subspace on which the corresponding representation exists is easily found. Let $v = \alpha(i_1 \otimes i_2) + \beta(i_1 \otimes j_2) + \Gamma(j_1 \otimes i_2) + \delta(j_1 \otimes j_2)$ be a typical vector. Then

$$P_E v = \begin{pmatrix} \frac{1}{2} & 0 & 0 & -\frac{1}{2} \\ 0 & \frac{1}{2} & \frac{1}{2} & 0 \\ 0 & \frac{1}{2} & \frac{1}{2} & 0 \\ -\frac{1}{2} & 0 & 0 & \frac{1}{2} \end{pmatrix} \begin{pmatrix} \alpha \\ \beta \\ \Gamma \\ \delta \end{pmatrix} = \frac{1}{2}\begin{pmatrix} \alpha - \delta \\ \beta + \gamma \\ \beta + \gamma \\ -(\alpha - \delta) \end{pmatrix}$$

$$\Rightarrow P_E = \frac{\alpha - \delta}{2}(i_1 \otimes i_2 - j_1 \otimes j_2) + \frac{\beta + \gamma}{2}(i_1 \otimes j_2 + j_1 \otimes i_2).$$

Thus $P_E$ projects onto the two-dimensional subspace spanned by $(i_1 \otimes i_2 - j_1 \otimes j_2)$ and $(i_1 \otimes j_2 + j_1 \otimes i_2)$. Likewise, one may show that $P_{A_1}$ projects onto the one-dimensional subspace spanned by

$$(i_1 \otimes i_2 + j_1 \otimes j_2)$$

and $P_{A_2}$ projects onto the one-dimensional subspace spanned by

$$(i_1 \otimes j_2 - j_1 \otimes i_2).$$

□

# Representations of Finite Groups

The preceding development illustrated the method by which a reducible representation can be decomposed into irreducible parts. In the remainder of this section, decomposition of tensor product of two irreducible representations is considered. Let $\Gamma^1$ and $\Gamma^2$ be two irreducible representations of a group $G$ and $\Gamma^1 \otimes \Gamma^2$ be their tensor product. The multiplicity of the unit representation of $G$ in $\Gamma^1 \otimes \Gamma^2$ may be calculated by using Equations 3.6.3 and 3.7.2. Since every group element has a unit character in the unit representation, the multiplicity of the unit representation is given by

$$\frac{1}{|G|} \sum_{g \in G} \chi^A(g) \overline{\chi^{\Gamma^1 \otimes \Gamma^2}(g)} = \frac{1}{|G|} \sum_{g \in G} \overline{\chi^{\Gamma^1} \chi^{\Gamma^2}}.$$

The right hand side of the above equation is equal to 1 if $\Gamma^1$ and $\Gamma^2$ are conjugate representations and 0 in all other cases (Equation 3.3.6). If the characters of the representation are real, then the unit representation is present only in the tensor product of the representation with itself. For now suppose $\Gamma$ is an irreducible representation of $G$ on a unitary space $V$ which is spanned by the basis $\{x_i\}_{i=1}^n$. Then the tensor product space $V \otimes V$ is of dimension $n^2$ and is spanned by basis states $\{x_i \otimes x_j\}_{i=1,j=1}^{n,n}$. Note that $x_i \otimes x_j$ is distinguished from $x_j \otimes x_i$. Apart from this particular choice of basis, another choice is sometimes useful. Consider the subspaces $(V \otimes V)^\sigma$ and $(V \otimes V)^\alpha$ (read symmetric and antisymmetric subspaces respectively). The symmetric space $(V \otimes V)^\sigma$ is spanned by basis states $x_i \otimes x_j + x_j \otimes x_i$ when $i \neq j$ and the states $x_i \otimes x_i$. Clearly the dimension of $(V \otimes V)^\sigma$ is $\frac{n(n+1)}{2}$. In a similar manner, the antisymmetric subspace $(V \otimes V)^\alpha$ spanned by basis states $x_i \otimes x_j - x_j \otimes x_i$ for $i \neq j$ is of dimension $\frac{n(n-1)}{2}$. Note that the dimensions of the symmetric and the antisymmetric subspaces add up to $n^2$ as they should. Let $g$ be some transformation in $G$. Consider the action of $g$ on a basis state in the antisymmetric subspace. In accordance with Equation 3.6.1, one has

$$g(x_i \otimes x_j - x_j \otimes x_i) \to g(x_i) \otimes g(x_j) - g(x_j) \otimes g(x_i)$$

$$g(x_i \otimes x_j - x_j \otimes x_i) \to \left(\sum_{k=1}^n [\Gamma(g)]_{ki} x_k\right) \otimes \left(\sum_{l=1}^n [\Gamma(g)]_{lj} x_l\right)$$

$$- \left(\sum_{l=1}^n [\Gamma(g)]_{lj} x_l\right) \otimes \left(\sum_{k=1}^n [\Gamma(g)]_{ki} x_k\right)$$

$$g(x_i \otimes x_j - x_j \otimes x_i) \to \sum_{k,l=1}^n [\Gamma(g)]_{ki} [\Gamma(g)]_{lj} (x_k \otimes x_l - x_l \otimes x_k)$$

$$g(x_i \otimes x_j - x_j \otimes x_i) \to \frac{1}{2} \sum_{k,l=1}^n \left([\Gamma(g)]_{ki} [\Gamma(g)]_{lj} - [\Gamma(g)]_{li} [\Gamma(g)]_{kj}\right) (x_k \otimes x_l - x_l \otimes x_k)$$

As is evident from the above map, the antisymmetric subspace gives a subrepresentation $(\Gamma \otimes \Gamma)^\alpha$ of the group $G$. The characters $\chi_\alpha^2$ of the antisymmetric representation can be easily obtained from the above map to be

$$\chi_\alpha^2(g) = \frac{1}{2} \sum_{i,j=1}^{n} \left( [\Gamma(g)]_{ii} [\Gamma(g)]_{jj} - [\Gamma(g)]_{ji} [\Gamma(g)]_{ij} \right)$$

$$\Rightarrow \chi_\alpha^2(g) = \frac{1}{2} \left( (\chi(g))^2 - \chi\left(g^2\right) \right). \tag{3.7.4}$$

The reader will do well to convince themselves that the symmetric subspace also gives a subrepresentation $(\Gamma \otimes \Gamma)^\sigma$ of $G$. The characters $\chi_\sigma^2(g)$ of the symmetric representation can be shown to be

$$\chi_\sigma^2(g) = \frac{1}{2} \left( (\chi(g))^2 + \chi\left(g^2\right) \right). \tag{3.7.5}$$

Finally, it may be noted that in the decomposition of the representation $\Gamma \otimes \Gamma$ as $(\Gamma \otimes \Gamma)^\sigma \oplus (\Gamma \otimes \Gamma)^\alpha$, the symmetric as well as the antisymmetric component may be a reducible representation of $\Gamma$.

## 3.8 Irreducible Representations of Direct Products

As has been noted previously, it is sometimes possible that a group is isomorphic to a direct product of two of its subgroups (e.g. $D_{nh} = D_n \times C_s$). The tensor product described in Section 3.6 allows the construction of all the irreducible representations of the direct product from the irreducible representations of the component groups. Let $G$ be isomorphic to the the direct product $H \times K$ of two groups $H$ and $K$. Then any element $g$ of $G$ can be expressed uniquely as

$$g = hk,$$

where $h$ is in $H$ and $k$ is in $K$. It is also known that $H$ and $K$ are normal subgroups in $G$ and elements of $H$ commute with elements of $K$. Let it be assumed that $h_1$ and $h_2$ are conjugate in $H$ and similarly for $k_1$ and $k_2$ in $K$. Then $h_1 k_1 = h h_2 h^{-1} k k_2 k^{-1}$ for some $h$ in $H$ and some $k$ in $K$. Because of the commutation it follows that $h_1 k_1 = (hk) h_2 k_2 (hk)^{-1}$, i.e., $h_1 k_1$ is conjugate to $h_2 k_2$ and by the same reasoning $h_1 k_1$, $h_1 k_2$, $h_2 k_1$ and $h_2 k_2$ are all conjugates in $G$. A conjugacy class of $G$ consists of all elements of the form $hk$ in which $h$'s come from one conjugacy class of $H$ and the $k$'s come from one conjugacy class of $K$. It follows that the number of conjugacy classes in $G$ is equal to the product of the number of conjugacy classes in $H$ and the number of conjugacy classes in $K$. For this reason the number of irreducible representations of $G$ is equal to the number of irreducible representations of $H$ times the number of irreducible representations of $K$. It is natural to expect that an irreducible representation of $G$ could be obtained by forming a tensor product of the an irreducible representation of $H$ with an irreducible representation of $K$. If $\Gamma_H$ is an irreducible representation of $H$ on a unitary space $V$ spanned by the basis $\{x_i\}_{i=1}^{n_1}$ whereas $\Gamma_K$ is an irreducible representation of $K$ on a unitary space $W$ spanned by the basis $\{y_i\}_{i=1}^{n_2}$, then a representation of $G$ on $V \otimes W$ is obtained from the map

$$(hk)(x_i \otimes y_j) \to h(x_i) \otimes k(y_j),$$

where $hk$ is an element of $G$. Proceeding as before, with maps such as above, it is not very difficult to show that the representations so obtained indeed satisfy Equations 3.3.6, 3.3.7, 3.3.8 and 3.3.9. Additionally, in analogy with Equation 3.6.3, in the current case one has

$$\chi^{\Gamma_H \otimes \Gamma_K}(hk) = \chi^{\Gamma_H}(h)\chi^{\Gamma_K}(k). \tag{3.8.1}$$

**Example 26.** The group $D_{2h}$ is a direct product of $D_2$ and $C_s$. The character tables for $D_2$ and $C_s$ follow:

| $D_2$ | $E$ | $C_2(z)$ | $C_2(y)$ | $C_2(x)$ | | |
|---|---|---|---|---|---|---|
| $A_1$ | 1 | 1 | 1 | 1 | | $x^2, y^2, z^2$ |
| $B_1$ | 1 | 1 | $-1$ | $-1$ | $z, R_z$ | $xy$ |
| $B_2$ | 1 | $-1$ | 1 | $-1$ | $y, R_y$ | $zx$ |
| $B_3$ | 1 | $-1$ | $-1$ | 1 | $x, R_x$ | $yz$ |

| $C_s$ | $E$ | $\sigma_h$ | | |
|---|---|---|---|---|
| $A'$ | 1 | 1 | $x, y, R_z$ | $x^2, y^2, z^2$ |
| $A''$ | 1 | $-1$ | $z, R_x, R_y$ | $yz, xz$ |

It is evident now that $D_{2h}$ has 8 conjugacy classes and 8 irreducible representations. Apart from all the conjugacy classes in $D_2$, there are other classes in $D_{2h}$. For example, $C_2(z)$ in $D_2$ multiplied with $\sigma_h$ in $C_s$ gives a class containing a single operation of inversion in $D_{2h}$, denoted by $I$. Similarly, multiplication of $C_2(y)$ with $\sigma_h$ is a class containing the single element corresponding to reflection in the $xz$ plane, and so on. The characters are various classes in $D_{2h}$ and are obtained from those of $D_2$ and $C_s$ by use of Equation 3.8.1. The names of various irreducible representations of $D_{2h}$, in accordance with the discussion in Section 3.4, are as mentioned in the character table:

| $D_{2h}$ | $E$ | $C_2(z)$ | $C_2(y)$ | $C_2(x)$ | $I$ | $\sigma_{xy}$ | $\sigma_{xz}$ | $\sigma_{yz}$ | | |
|---|---|---|---|---|---|---|---|---|---|---|
| $A_g$  | 1 | 1  | 1  | 1  | 1  | 1  | 1  | 1  |       | $x^2, y^2, z^2$ |
| $B_{1g}$ | 1 | 1  | -1 | -1 | 1  | 1  | -1 | -1 | $R_z$ | $xy$ |
| $B_{2g}$ | 1 | -1 | 1  | -1 | 1  | -1 | 1  | -1 | $R_y$ | $zx$ |
| $B_{3g}$ | 1 | -1 | -1 | 1  | 1  | -1 | -1 | 1  | $R_x$ | $yz$ |
| $A_u$  | 1 | 1  | 1  | 1  | -1 | -1 | -1 | -1 |       |      |
| $B_{1u}$ | 1 | 1  | -1 | -1 | -1 | -1 | 1  | 1  | $z$ |    |
| $B_{2u}$ | 1 | -1 | 1  | -1 | -1 | 1  | -1 | 1  | $y$ |    |
| $B_{3u}$ | 1 | -1 | -1 | 1  | -1 | 1  | 1  | -1 | $x$ |    |

$\square$

## 3.9 Induced Representations

Let $G$ be a group and $\Gamma$ a representation of $G$ on a vector space $V$. For a subgroup $H$ of $G$, the matrices $\Gamma(h)$ for all $h \in H$ together form a representation of $H$ on $V$. A representation of $H$ obtained by restricting a representation of $G$ to $H$ is denoted by $\Gamma_*$ in this section. The reverse problem of obtaining a representation of $G$ from a representation $\Theta$ of $H$ is considered here. Such a representation of $G$ is said to be induced by the representation $\Theta$ of $H$ and is denoted by $\Theta^*$ in this section.

The first step in the construction of $\Theta^*$ is to realize a representation of $G$ on the set of cosets of $H$ in $G$. Let $\{x_i\}_{i=1}^n$ be a set of representatives, one from each of the distinct cosets of $H$ in $G$. Then one has $G = \cup x_i H$. If $g \in G$, then $g x_i H$ is again a coset and therefore equal to $x_k H$ for some $x_k$. Because $\{x_i\}_{i=1}^n$ are representatives of distinct cosets, $g x_i H \neq g x_j H$ unless $i = j$. In other words, left multiplication by $g$ permutes the cosets of $H$ among themselves. Every $g \in G$ in this way can be identified with a permutation $\tau_g$ in $\mathfrak{S}(n)$, the symmetric group on the $n$ letters $\{x_i\}_{i=1}^n$. It follows from

$$\tau_{g_1 g_2}(x_i H) = g_1 g_2 (x_i H) = g_1 (g_2 x_i H)$$

$$\Rightarrow \tau_{g_1 g_2}(x_i H) = \tau_{g_1} \tau_{g_2}(x_i H)$$

that $\tau$ is a representation of $G$ into $\mathfrak{S}(n)$, i.e., $\tau : G \to \mathfrak{S}(n)$ is a homomorphism. One notices that the regular representation described in the Section 3.5 corresponds to the case when $H$ is the trivial subgroup $\{E\}$ of $G$.

**Example 27.** Consider the group $\mathfrak{S}(3)$ of Example 10 in Chapter 1. The subgroup $H = \{e, a\}$ has 3 cosets $eH = \{e, a\}$, $bH = \{b, ba\}$ and $b^2 H = \{b^2, b^2 a\}$. Consider the action of the element $ab \in \mathfrak{S}(3)$ on the cosets. Evidently

$$ab : eH \to b^2 H, \quad ab : bH \to bH, \quad ab : b^2 H \to eH.$$

In matrix form, the above action can be written as

$$\tau_{ab} = \begin{pmatrix} 0 & 0 & 1 \\ 0 & 1 & 0 \\ 1 & 0 & 0 \end{pmatrix}.$$

The representation for other elements can be similarly worked out.

$$\tau_e = \begin{pmatrix} 1 & 0 & 0 \\ 0 & 1 & 0 \\ 0 & 0 & 1 \end{pmatrix}, \tau_a = \begin{pmatrix} 1 & 0 & 0 \\ 0 & 0 & 1 \\ 0 & 1 & 0 \end{pmatrix}, \tau_b = \begin{pmatrix} 0 & 0 & 1 \\ 1 & 0 & 0 \\ 0 & 1 & 0 \end{pmatrix},$$

$$\tau_{b^2} = \begin{pmatrix} 0 & 1 & 0 \\ 0 & 0 & 1 \\ 1 & 0 & 0 \end{pmatrix}, \tau_{ab^2} = \begin{pmatrix} 0 & 1 & 0 \\ 1 & 0 & 0 \\ 0 & 0 & 1 \end{pmatrix}.$$

$\square$

As previously, let $\Theta$ be a representation of the subgroup $H$ over a vector space $W$ of dimension $m$. $\Theta$ is extended to $\Theta^*$ by $\tau$ action of $G$ on the coset space of $H$. The $\tau$ representation of $G$ consists of permutation matrices. Such matrices have a single 1 as the only non-zero entry in each row and each column as is evident from the above example. For $g \in G$, the $(i,j)$'th entry in $\tau_g$ is equal to 1 if and only if $x_i^{-1} g x_j \in H$. Define matrices $g^{ij}$ as

$$g^{ij} = \begin{cases} \Theta(x_i^{-1} g x_j) & \text{if } x_i^{-1} g x_j \in H \\ 0_{m \times m} & \text{otherwise.} \end{cases} \tag{3.9.1}$$

Here $0_{m \times m}$ is the $m \times m$ matrix all of whose entries are equal to 0. The matrix $\Theta^*(g)$ in the induced representation is now given in terms of matrix blocks $g^{ij}$ as

$$\Theta^*(g) = \begin{pmatrix} g^{11} & \cdots & g^{1n} \\ \vdots & \ddots & \vdots \\ g^{n1} & \cdots & g^{nn} \end{pmatrix}. \tag{3.9.2}$$

Evidently, the order of $\Theta^*(g)$ is $mn \times mn$. Because of the form of $\tau$, only one matrix block is non-zero in any row or any column of matrix blocks in $\Theta^*(g)$. It remains to be verified that $\Theta^*$ is a representation of $G$. For $g_1, g_2 \in G$ the $(i,j)$'th matrix block of the product matrix $\Theta^*(g_1)\Theta^*(g_2)$ is given by

$$[\Theta^*(g_1)\Theta^*(g_2)]^{ij} = \sum_{k=1}^{n} (g_1)^{ik} (g_2)^{kj}.$$

The matrix in the right hand side of the above can be non-zero only when $(g_1)^{ik}$ and $(g_2)^{kj}$ are non-zero for the same $k = k_0$. This happens when both $x_i^{-1} g_1 x_{k_0}$ and $x_{k_0}^{-1} g_2 x_j$

are in $H$. Then $(g_1)^{ik_0} = \Theta(x_i^{-1}g_1x_{k_0})$, $(g_2)^{k_0j} = \Theta(x_{k_0}^{-1}g_2x_j)$ and one has

$$[\Theta^*(g_1)\Theta^*(g_2)]^{ij} = \Theta(x_i^{-1}g_1x_{k_0})\Theta(x_{k_0}^{-1}g_2x_j)$$

$$\Rightarrow [\Theta^*(g_1)\Theta^*(g_2)]^{ij} = [\Theta^*(g_1g_2)]^{ij}. \quad (3.9.3)$$

Also note that $g_2x_j \in x_kH$ for some $x_k$ and if $x_i^{-1}g_1g_2x_j \in H$, then both $x_i^{-1}g_1x_k$ and $x_k^{-1}g_2x_j$ are in $H$. Hence the corresponding non-zero blocks of $\Theta^*(g_1)\Theta^*(g_2)$ and $\Theta^*(g_1g_2)$ are equal. It is similarly shown that the same holds for the zero blocks and Equation 3.9.3 holds generally. It follows that $\Theta^*$ is indeed a representation of $G$.

**Example 28.** Continuing with the previous example, a representation of $H$ of degree 2 is

$$\Theta(e) = \begin{pmatrix} 1 & 0 \\ 0 & 1 \end{pmatrix}; \Theta(a) = \begin{pmatrix} 0 & 1 \\ 1 & 0 \end{pmatrix}.$$

The reader should follow the above description to obtain $\Theta^*$. A couple of representing matrices are

$$\Theta^*(ab) = \begin{pmatrix} 0 & 0 & 0 & 0 & 0 & 1 \\ 0 & 0 & 0 & 0 & 1 & 0 \\ 0 & 0 & 0 & 1 & 0 & 0 \\ 0 & 0 & 1 & 0 & 0 & 0 \\ 0 & 1 & 0 & 0 & 0 & 0 \\ 1 & 0 & 0 & 0 & 0 & 0 \end{pmatrix}; \Theta^*(b^2) = \begin{pmatrix} 0 & 0 & 1 & 0 & 0 & 0 \\ 0 & 0 & 0 & 1 & 0 & 0 \\ 0 & 0 & 0 & 0 & 1 & 0 \\ 0 & 0 & 0 & 0 & 0 & 1 \\ 1 & 0 & 0 & 0 & 0 & 0 \\ 0 & 1 & 0 & 0 & 0 & 0 \end{pmatrix}. \quad \square$$

If $\chi$ is the character of $\Theta$ and $\chi^*$ that of $\Theta^*$, then it follows from Equation 3.9.2 that $\chi^*(g) = \sum_{i=1}^n \text{tr } g^{ii}$. By use of Equation 3.9.1, one has

$$\chi^*(g) = \sum_{i=1,\, x_i^{-1}gx_i \in H}^n \chi(x_i^{-1}gx_i). \quad (3.9.4)$$

For $x_i^{-1}gx_i \in H$, one also has $\chi(x_i^{-1}gx_i) = \chi(h^{-1}x_i^{-1}gx_ih)\ \forall\ h \in H$. Employing this fact, the reader can easily establish

$$\chi^*(g) = \frac{1}{|H|} \sum_{t \in G,\, t^{-1}gt \in H} \chi(t^{-1}gt). \quad (3.9.5)$$

Suppose now that $\Gamma$ is an irreducible representation of $G$ and $\Theta$ an irreducible representation of $H$, a subgroup of $G$. In general, $\Gamma_*$ and $\Theta^*$ are reducible representations of $H$ and $G$ respectively. The multiplicity $m_*$ of $\Theta$ in $\Gamma_*$ is given by

$$m_* = \frac{1}{|H|} \sum_{h \in H} \chi^\Gamma(h)\chi^\Theta(h^{-1}),$$

where use has been made of the facts $\Gamma_*(h) = \Gamma(h) \ \forall \ h \in H$ and $\overline{\chi^\Theta(h)} = \chi^\Theta(h^{-1})$. The multiplicity $m^*$ of $\Gamma$ in $\Theta^*$ is given by

$$m^* = \frac{1}{|G|} \sum_{g \in G} \chi^{\Theta^*}(g^{-1}) \chi^\Gamma(g)$$

$$\Rightarrow m^* = \frac{1}{|G|} \sum_{g \in G} \chi^\Gamma(g) \left( \sum_{i=1, x_i^{-1}g^{-1}x_i \in H}^n \chi^\Theta(x_i^{-1}g^{-1}x_i) \right).$$

Rearranging the summations and using the fact that $\chi^\Gamma(g) = \chi^\Gamma(x_i^{-1}gx_i)$, one has

$$m^* = \frac{1}{|G|} \sum_{i=1}^n \sum_{g \in G, \ x_i^{-1}g^{-1}x_i \in H} \chi^\Gamma\left(x_i^{-1}gx_i\right) \chi^\Theta\left(x_i^{-1}g^{-1}x_i\right).$$

The conjugation $x_i^{-1}Gx_i$ is an automorphism of $G$. The second summation is therefore over exactly all the elements of $H$. Since there are $|G|/|H|$ representatives, it follows that

$$m^* = \frac{1}{|G|} \frac{|G|}{|H|} \sum_{h \in H} \chi^\Gamma(h) \chi^\Theta(h^{-1}) = m_*. \tag{3.9.6}$$

The above result is called the *Frobenius reciprocity theorem*.

Assuming again $H$ to be a subgroup of $G$, let $\Theta_1$ and $\Theta_2$ be representations of $H$ over vector spaces of $W_1$ and $W_2$. Let the degrees of these representations be $n_1$ and $n_2$. Then in the representation $\Theta_1 \oplus \Theta_2$ over $W_1 \oplus W_2$, one has for $h \in H$

$$(\Theta_1 \oplus \Theta_2)(h) = \begin{pmatrix} \Theta_1(h) & 0_{n_1 \times n_2} \\ 0_{n_2 \times n_1} & \Theta_2(h) \end{pmatrix}.$$

The matrices in the representation $(\Theta_1 \oplus \Theta_2)^*$ would have blocks such as above in places where 1 appears in the permutation representation of $G$ on the coset space of $H$ and $0_{(n_1+n_2) \times (n_1+n_2)}$ blocks elsewhere. Such a matrix can be written as a sum of two matrices which operate on complimentary space isomorphic to $W_1$ and $W_2$. Therefore, one has

$$(\Theta_1 \oplus \Theta_2)^* = \Theta_1^* \oplus \Theta_2^*. \tag{3.9.7}$$

## Exercises

1. Show that all invertible $n \times n$ matrices with complex entries form a group. Also show that the set of matrices whose determinant is of unit modulus is a subgroup.

2. Show that two matrices related by a similarity transformation have the same trace.

3. Using the orthogonality relations, write the character table of the group $C_4$ and indicate the irreducible representations using Mulliken's symbols. Write down the matrix form of the regular representation of $C_4$.

4. Show that the characters of a finite cyclic group in any irreducible unitary representation are unimodular.

5. In a certain representation $\Gamma$ of $C_{3v}$, it is known that $\chi^\Gamma(E) = 7$, $\chi^\Gamma(C_3) = 1$ and $\chi^\Gamma(\sigma_v) = -3$. Determine the multiplicities of various irreducible representations of $C_{3v}$ in $\Gamma$.

6. Obtain the maximal point group symmetry for a particle in a two-dimensional box. Write down the character table for this group along with the basis elements (consider the action of group elements on the coordinate space). Further figure out which stationary states of the particle give the irreducible representations of the group.

7. Construct the character table of the group $C_{4v}$. Determine the irreducible representations present in the tensor product $E \otimes E$ where $E$ is the two-dimensional irreducible representation of $C_{4v}$.

8. For the symmetric group $\mathfrak{S}(3)$, determine the characters of all the irreducible representations. Give a complete reducible $3 \times 3$ representation $\Gamma$ of $\mathfrak{S}(3)$ on a three-dimensional vector space $V$ spanned by $(1, 0, 0)$, $(0, 1, 0)$ and $(0, 0, 1)$. Determine the multiplicity of all the irreducible representations in $\Gamma$. Using the projection operator on $\Gamma$, project out the subspaces of $V$ on which $\Gamma$ is irreducible.

9. Prove Equation 3.7.5.

10. Construct the character table for the point group $D_{3d}$. Consider $D_{3d}$ as a direct product of two of its subgroups.

11. Complete the proof of Equation 3.9.5.

12. Let $G$ be a finite group and $H$ a subgroup of $G$. Show that the regular representation of $H$ induces the regular representation $G$.

13. If $H$ is a subgroup of $G$, then show that any irreducible representation of $G$ is a component in the representation induced by some irreducible representation of $H$. The result of the previous exercise is useful.

14. In the notation of the last section of the chapter, show that $\Gamma \otimes \Theta^* = (\Gamma_* \otimes \Theta)^*$.

# 4

# Elementary Applications

In this chapter, we will focus on solving problems in physics using the group theory representations extensively discussed in the previous chapter. The character tables for any discrete group will be the primary tool to apply in these physical problems. Particular attention will be on atomic and nuclear systems obeying the postulates of quantum mechanics.

For instance, a physical system possessing a discrete group symmetry like $C_{4v}$ may get disturbed by impurities or defects. Such a disturbance could result in a system with a residual symmetry like $C_2$. In fact, we will see how these effects are reflected in energy spectrum observed experimentally which can be validated using the group representation theory tools.

We hope the readers will appreciate how the formal mathematical steps provide meaning to results observed in the physical systems. We plan to present several examples in atomic, nuclear and particle physics which will highlight the elegance of group theory tools.

## 4.1 General Considerations

We will briefly review the salient features of quantum mechanics which describe the symmetries possessed by microscopic systems. In fact, this review will be useful for validating these quantum mechanical results applying group theory tools.

For a closed quantum mechanical system, the eigenstates $\psi$ of the system are solutions of the time independent Schrödinger's equation:

$$\mathbf{H}\psi = E\psi, \tag{4.1.1}$$

where H denotes Hamiltonian of the system and $E$ is the energy when the system is in the stationary state $\psi$. The Hamiltonian H of the system is a self-adjoint operator that acts on the Hilbert space of system states. Equation 4.1.1 essentially states that the action of the Hamiltonian on a stationary state leaves the state invariant apart from a constant factor of $E$. The eigenstate $\psi$ can be always taken to be a unit vector in the Hilbert space. All vectors which are complex multiples of the same unit vector are assumed to be physically equivalent. In this sense, $\psi$ represents that normalized state of the system. In the notation of Section 3.1, this fact is represented by

$$(\psi, \psi) = 1. \tag{4.1.2}$$

Let O be a unitary operator which commutes with H. By the fact that O and H commute, it is implied that the order in which O and H act on a state $\psi$ is immaterial, i.e., HO $\psi$ = OH$\psi$. If $\psi$ is a stationary state of the system satisfying Equation 4.1.1, then one has

$$\mathbf{H}(\mathbf{O}\psi) = E\mathbf{O}\psi.$$

In other words, if $\psi$ is a stationary state of the system, then so is O$\psi$. In general, O$\psi$ would differ from $\psi$ not merely by a constant complex multiple. In this circumstance, the energy level $E$ is said to be *degenerate*.

There can be a set of unitary operators $\{O_1, O_2, \cdots\}$ commuting with H. By a similar application of these operators to the Schrödinger equation 4.1.1, we deduce energy $E$ to be same for states $\{O_i\psi\}$ which may be linearly independent to $\psi$. Suppose this set of unitary operators form a group $G$. Then we interpret such a quantum mechanical system to possess the group symmetry $G$.

In Section 3.2, it was noted how the elements of a group $G$ may act on a vector space and give a representation of $G$ on that vector space. Suppose that the group $G$ is a symmetrical group of the system under consideration. Such symmetry transformations would leave the form of the Schrödinger equation invariant. We denote the representation of unitary operator $O_i$ as $\Gamma(g_i)$, corresponding to any group element $g_i \in G$, acting on the Hilbert space. Particularly the elements of the set $\{\Gamma(g_i)\}$ commutes with the Hamiltonian of the system possessing group symmetry $G$. The energy of a stationary state cannot change due to action of $\Gamma(g_i)$. Thus the subspace of degenerate states corresponding to some energy level $E$ gives a representation of $G$. As this subrepresentation is unitary, it is completely reducible. It is then possible to choose the degenerate eigenstates corresponding to an energy level $E$ so that they transform according to irreducible representations of $G$. If a particular irreducible representation occurs only once in the representation of $G$ on the full Hilbert space, then all the basis states of this representation are also all the degenerate eigenstates corresponding to some energy $E$. It may so happen that the degenerate levels of another energy $E'$ give a reducible representation of $G$. This happens if the Hamiltonian of the system has a higher order symmetry than $G$, a circumstance we will assume does not occur in the subsequent discussion. To sum up, it would be assumed that the degenerate states corresponding to an energy level can be so chosen that they give an irreducible representation of the symmetry group of the system.

# Elementary Applications

**Example 29.** For a one-dimensional harmonic oscillator given by the Hamiltonian

$$\mathbf{H} = -\frac{\hbar^2}{2m}\frac{d^2}{dx^2} + \frac{1}{2}m\omega^2,$$

determine the group symmetry and deduce whether the energy eigenstates are degenerate or non-degenerate.

Observe that the inversion transformation $g : x \to -x$, where $g \in C_i$, commutes with **H**. Hence $C_i$ is the group symmetry. Recall that the character table of abelian groups contain only one-dimensional irreducible representations. Hence, the energy eigenstates of the harmonic oscillator possessing $C_i$ group symmetry must be non-degenerate.

We have illustrated the power of the group theory tool to deduce the non-degenerate nature of energy eigenstates without solving the Schrödinger equation. Further, the stationary states

$$\psi(x) \propto \Gamma(g)\psi(x),$$

where $\Gamma(g)$ denotes the representation of the group element acting on the Hibert space. By using the projection method (see Section.3.7), the basis states of group $C_i$ can be shown to be either

$$\text{odd}: \psi_0(x) = -\psi_0(-x) \quad \text{or even}: \psi_e(x) = \psi_e(-x).$$

Thus group theory arguments have justified that the stationary states are non-degenerate states which are either $\psi_0(x)$ or $\psi_e(x)$. However, to determine the explicit form of stationary states and the energy eigenvalues, we have to solve the Schrödinger equation. □

We will now look at another system with degenerate energy levels.

**Example 30.** Consider a particle of mass $m$ moving in two dimensions subjected to potential $V(x, y)$ which is zero inside a square region $-a \leq (x, y) \leq a$ and $\infty$ elsewhere. This is the familiar example in quantum mechanics known as particle in a square box whose energy eigenstates are either non-degenerate or two-fold degenerate. Can we explain the degenerate energy level using the discrete symmetry possessed by the system ?

The Hamiltonian describing such a system is

$$\mathbf{H} = -\frac{\hbar^2}{2m}\left(\frac{\partial^2}{\partial x^2} + \frac{\partial^2}{\partial y^2}\right) + V(x, y),$$

whose stationary states can be determined by solving the Schrodinger Equation 4.1.1. We quote the result which is extensively discussed in any quantum mechanics textbook. The energy eigenstates and eigenvalues are

$$\psi_{(n_1,n_2)}(x, y) = \left(\frac{1}{a}\right)\sin\left(\frac{n_1\pi x}{2a}\right)\sin\left(\frac{n_2\pi y}{2a}\right); \quad E_{(n_1,n_2)} = (n_1^2 + n_2^2)\frac{\pi^2\hbar^2}{8ma^2}.$$

Clearly, $\psi_{(n_1,n_2)}(x, y)$ and $\psi_{(n_2,n_1)}(x, y)$ are linearly independent states sharing the same energy eigenvalues.

The rotation operation by angle $\pi/2$ about z-axis, which transforms $(x, y) \to (y, -x)$, as well as mirror reflection $\sigma_v$ commutes with the Hamiltonian. Hence, the group symmetry possessed by this system is $C_{4v}$. We observe the character table of $C_{4v}$ has one-dimensional as well as two-dimensional irreducible representations. Hence, we can claim that the energy states are either non-degenerate or two-fold degenerate. It is appropriate to mention that besides the group symmetry accounting for degenerate energy levels, there could be accidental degeneracy. For example, the stationary states with $(n_1, n_2) = (5,5), (7, 1), (1, 7)$ share the same energy eigenvalues where the accidental degenerate state $\psi_{(5,5)}(x, y)$ cannot be deduced using group theory tools.

Suppose the square region, where potential is $V(x,y) = 0$, gets slightly modified to a rectangular region due to mild disturbance. We know from quantum mechanics that the two-fold degenerate levels in a square box split into two non-degenerate levels in such a rectangular box of arbitrary length and breath (we neglect accidental degeneracy which can arise due to suitable ratio of the sides of the rectangle). From the group theory point of view, we can say that the two-fold degenerate energy levels show splitting due to disturbance. In the following section, we will understand how disturbance or perturbation can split the degenerate energy levels using group theory tools. □

## 4.2 Level Splitting under Perturbation

For a system described by H possessing symmetry group G, Equation 4.1.1 will give the stationary states. Let the degenerate eigenstates $\{\psi_i\}_{i=1}^{n}$ give an irreducible unitary representation $\Gamma$ of the symmetry group $G$ of the system. The characters of $G$ in $\Gamma$ are of course assumed known. On application of a perturbation $\mathbf{H_1}$ to the system, the Schrödinger equation takes the form

$$(\mathbf{H} + \mathbf{H_1})\psi = E\psi.$$

If the perturbation is assumed to be small, the stationary states and the energy levels of the perturbed system are expected to differ only slightly from those of the unperturbed system. It is known from perturbation theory methods in quantum mechanics that a small perturbation usually lifts the degeneracy in a set of degenerate levels such as $\{\psi_i\}_{i=1}^{n}$. While a complete calculation of the perturbed levels is made possible only through application of perturbation methods to the known form of $H_1$, group theory allows one to deduce the extent to which the degeneracy gets lifted from the form of $H_1$ in some cases.

Suppose K is the maximal group of spatial transformations which commutes with the perturbation. If K is a subgroup of G then the perturbation $H_1$ has a lower order symmetry than the unperturbed Hamiltonian H. Consequently, the complete Hamiltonian of the perturbed system also has its symmetry group as K. Since $\Gamma$ is a representation of G, it is also a representation of the subgroup K. As $\Gamma$ is unitary, it is

# Elementary Applications 63

now possible to completely reduce $\Gamma$ as a direct sum of irreducible representations of $K$. These irreducible representations of $K$ correspond to different energy eigenvalues of the perturbed system in accordance with our assumption. It is immediate that Equations 3.7.1 and 3.7.2 are directly applicable in this case.

Using the above arguments, we can now revisit Example 30 where $V(x, y) = 0$ in a rectangular region due to perturbation. The group symmetry possessed by the particle in a rectangular box system is no longer $G = C_{4v}$ but $K = C_{2v}$ which is abelian. Hence, any two-dimensional irreducible representation $\Gamma \in C_{4v}$ will become the direct sum of irreducible representation: $\oplus \Gamma^\alpha \in C_{2v}$ which are one-dimensional representations. Therefore all the two-fold degenerate levels of the square box will split into non-degenerate levels (ignoring accidental degeneracy). We will elaborate another example which will illustrate the splitting of energy levels using the group theory tools.

**Example 31.** Let the unperturbed Hamiltonian of a system have $T_d$ symmetry (Chapter 2, Section 2.3). Suppose a small perturbation breaks the symmetry to the subgroup $C_{3v}$. For reference, character tables for $T_d$ and $C_{3v}$ are given below. Let us determine the splitting of energy level corresponding to a representation of degree 3 (say $F_1$) in $T_d$.

| $T_d$ | E | $8C_2$ | $3C_2$ | $6\sigma_d$ | $6S_4$ |
|---|---|---|---|---|---|
| $A_1$ | 1 | 1 | 1 | 1 | 1 |
| $A_2$ | 1 | 1 | 1 | −1 | −1 |
| E | 2 | −1 | 2 | 0 | 0 |
| $F_1$ | 3 | 0 | −1 | 1 | −1 |
| $F_2$ | 3 | 0 | −1 | −1 | 1 |

| $C_{3v}$ | E | $2C_3$ | $3\sigma_v$ |
|---|---|---|---|
| $A_1$ | 1 | 1 | 1 |
| $A_2$ | 1 | 1 | −1 |
| E | 2 | −1 | 0 |

Some observations are in order before calculating the split. The permutation representation of $T_d$ is the symmetric group $\mathfrak{S}(4)$. Similarly, the permutation representation of $C_{3v}$ is the symmetric group $\mathfrak{S}(3)$. Since $\mathfrak{S}(3)$ is a subgroup of $\mathfrak{S}(4)$, it follows that $C_{3v}$ is a subgroup of $T_d$. It may be recalled here that $\mathfrak{S}(4) = V_N \rtimes \mathfrak{S}(3)$. In the point group $T_d$, the normal subgroup corresponding to $V_N$ (the Klein-4 normal subgroup in $\mathfrak{S}(4)$) consists of the identity transformation along with the three $C_2$ rotations which interchange pairs of vertices of the tetrahedron. In the isomorphism

$$\frac{T_d}{V_N} \cong C_{3v},$$

the $8C_3$ class in $T_d$ is mapped to $2C_3$ class in $C_{3v}$, the classes $6\sigma_d$ and $6S_4$ in $T_d$ are both mapped to the class $3\sigma_v$ in $C_{3v}$ while $V_N$ being the kernel of the homomorphism is mapped to the identity class in $C_{3v}$. The characters of the representations of $C_{3v}$ can be extended in this case to the characters of representations of the same degrees in $T_d$ by noting the fact that all representations of the kernel $V_N$ are of degree 1 since $V_N$ is an abelian group. For example, consider the representation E. The characters of the identity class and the $3C_2$ class in $T_d$ are both equal to 2, the character of the identity class of $C_{3v}$ (to which both the classes of $T_d$ are mapped in the homomorphism). Similarly one may obtain the characters of other representations of $T_d$ from those of $C_{3v}$. However, since $C_{3v}$ has no irreducible representations of degree 3, the characters of the representations $F_1$ and $F_2$ of $T_d$ would have to be obtained via application of the orthogonality relations on the characters.

Returning to the splitting of the degeneracy of $F_1$ under a $C_{3v}$ perturbation, the basis states of the representation $F_1$ of $T_d$ give a reducible representation of $C_{3v}$. In this reducible representation, the characters of the various classes of $C_{3v}$ can be read directly from the characters of $F_1$ by considering the homomorphism mentioned above and orthogonality relations:

| $C_{3v}$ | $E$ | $2C_3$ | $3\sigma_v$ |
|---|---|---|---|
| $\Gamma$ | 3 | 0 | 1 |

Now apply Equation 3.7.2 (which we note once again)

$$m_\alpha = \frac{1}{|G|} \sum_{g \in G} \chi^\Gamma(g) \overline{\chi^{\Gamma^\alpha}(g)}.$$

In the above, $\Gamma$ is the reducible representation of $C_{3v}$ effected by the basis states of the $F_1$ representation of $T_d$ and $\Gamma^\alpha$ is any of the irreducible representations of $C_{3v}$. Upon carrying out the routine calculations, it is found that

$$\Gamma = A_1 \oplus E.$$

The above indicates that the perturbation splits the triply degenerate energy level of the original Hamiltonian into two levels, one singlet and the other one doubly degenerate. □

Stationary states are the stable energy levels in which the system stays as long as there is no external interactions. Experimentalists in the laboratory measure the frequencies $\{v_i\}'s$ of spectral lines emitted whenever an atom undergoes a transition from one energy level to another energy level due to interactions. The observed frequencies are characteristic of an atom and also depend on the nature of interaction. This brings us to topic of *selection rules* where group theory tools can deduce whether an atomic transition is allowed or disallowed due to such interactions.

## 4.3 Selection Rules

In quantum mechanics, physical quantities are represented by self-adjoint operators acting on the Hilbert space of system states. The Hamiltonian of the system is itself an example of such an operator which is useful to determine the energy eigenvalues $E_i$ and their corresponding stationary states $\psi_i$ solving Schrödinger equation. The stationary states (indexed by a complete set of quantum numbers) form a basis for the Hilbert space. There are other linear operators corresponding to physical quantities like position, momentum, angular momentum and so forth, which can be expressed

# Elementary Applications

in a matrix form in this basis. Let $\mathcal{O}$ represent a self-adjoint operator corresponding to a physical quantity. The matrix of $\mathcal{O}$ in the basis of stationary states is then given by

$$\mathcal{O}_{mn} = \int \psi_m^* \mathcal{O} \psi_n = (\psi_m, \mathcal{O}\psi_n), \tag{4.3.1}$$

where $\{\psi_i\}$ are the stationary states indexed by a complete set of quantum numbers. The integration in the equation above is carried out over the complete configuration space of the system. The non-zero matrix element $\mathcal{O}_{mn}$ implies that the transition from energy level $E_n$ to another energy level $E_m$ due to interaction represented by operator $\mathcal{O}$ is allowed. Instead of working this matrix element by integrating Equation 4.3.1, we would like to exploit the power of group theory to deduce whether $\mathcal{O}_{mn}$ is zero or non-zero.

Suppose the system possesses group symmetry $G$ corresponding to spatial transformation. We expect the above matrix element $\mathcal{O}_{mn}$ to be unchanged when any group element $g \in G$ acts on the states $\psi_m, \psi_n$ as well as on the operator $\mathcal{O}$. In fact, one can deduce which of the matrix elements $\mathcal{O}_{mn}$ vanishes by applying the principal ideas of group theory as follows.

For the system described by Equation 4.1.1 having $G$ as its non-trivial symmetry group, we know from Section 4.1 that the degenerate states of same energy belong to an irreducible representation of $G$. Consider the integral

$$\int \psi_i^\alpha$$

where $\psi_i^\alpha$ is a basis state in some irreducible representation $\Gamma^\alpha$ of the $G$. The action of any group element on $\psi_i^\alpha$ transforms into a linear combination of the basis states of same energy to which $\psi_i^\alpha$ belongs. However, a mere spatial transformation which brings the system to an identical configuration should not change the value of the integral $\int \psi_i^\alpha$. In other words it is expected that

$$\int \psi_i^\alpha = \sum_{k=1}^d [\Gamma^\alpha(g)]_{ki} \int \psi_k^\alpha,$$

where $d$ is the degree of the representation $\Gamma^\alpha$. Also, the above must be true for any transformation $g$ in $G$. Adding the expressions for all transformations $g$ in $G$, one has

$$\int \psi_i^\alpha = \frac{1}{|G|} \sum_{k=1}^d \left( \sum_{g \in G} [\Gamma^\alpha(g)]_{ki} \right) \int \psi_k^\alpha.$$

If $\Gamma^\alpha$ is not the trivial representation, then it is easy to verify using Equation 3.3.5 that $(\sum_{g \in G} [\Gamma^\alpha(g)]_{ki})$ vanishes. Therefore, if $\psi_i^\alpha$ is a basis state in some non-trivial irreducible unitary representation $\Gamma^\alpha$ of the $G$ then

$$\int \psi_i^\alpha = 0. \tag{4.3.2}$$

Returning to Equation 4.3.1, let $\psi_m^f$ and $\psi_n^i$ be eigenstates in two distinct irreducible representations $\Gamma^f$ and $\Gamma^i$ respectively, of the symmetry group $G$ of the system. The operator $\mathcal{O}$ is in general some tensor (for example, a dipole moment or a quadrupole moment) which has a certain number of independent components. It will be assumed that these independent components themselves are transformed into a linear combination of each other under the spatial transformations effected by the elements of $G$. It is then possible to treat those independent components of $\mathcal{O}$ as a basis for irreducible representation of $G$. For instance, see Example 32 for electric dipole moment components. Formally for any operator $\mathcal{O}$, its representation of $G$ is denoted as $\Gamma^{\mathcal{O}}$. By suitable linear combinations, it is always possible to choose the eigenfunctions of the stationary states to be real and $\psi_m$ in Equation 4.3.1 can be taken as real. Then the states $\psi_m^{f*} \mathcal{O} i$ would be transformed by $G$ according to the states of the tensor product $\Gamma^f \otimes \Gamma^{\mathcal{O}} \otimes \Gamma^i$. The representation $\Gamma^f \otimes \Gamma^{\mathcal{O}} \otimes \Gamma^i$ can obviously be decomposed into a direct sum of irreducible representations of $G$ and the multiplicities of these irreducible representations can be found by using Equation 3.7.2. The evaluation of $\mathcal{O}_{mn}$ reduces to the calculation of the sum of integrals of the form $\int \psi_k^\alpha$ where the $\psi_k^\alpha$'s transform in accordance with some irreducible representation of $G$. That all such integrals vanish has already been established unless of course some of the $\psi_k^\alpha$'s transform in accordance with the trivial representation. Therefore, the matrix elements of $\mathcal{O}$ vanish unless the trivial representation is present in the decomposition of $\Gamma^j \otimes \Gamma^{\mathcal{O}} \otimes \Gamma^i$. Alternatively, when the tensor product $\Gamma^{\mathcal{O}} \otimes \Gamma^i$ is decomposed into irreducible components, then these irreducible components are the possible values of $\Gamma^f$ for which the transition elements may be non-zero. This is a very convenient way to decide which matrix elements vanish as against a full-fledged evaluation of the integral. The exact values of the non-vanishing matrix elements would of course have to be calculated by integration. The implementation of this formalism is detailed in Examples 32 and 33.

The aforementioned method for calculation of $\mathcal{O}_{mn}$ works when $\psi_m$ and $\psi_n$ belong to different energy levels of the system. In order to consider transition elements which are diagonal with respect to energy, we make a small digression to recall an important fact related to the solution of the time dependent Schrödinger equation

$$i\frac{\partial}{\partial t}\Psi = H\Psi.$$

Let the Hamiltonian H of the system described by the time dependent wave function $\Psi$ be a time independent real operator. If the state $\Psi$ of the system develops in accordance with the above equation, then the conjugates of quantities on both sides of the equation should be always equal, i.e.,

$$-i\frac{\partial}{\partial t}\overline{\Psi} = H\overline{\Psi}$$

$$\Rightarrow i\frac{\partial}{\partial(-t)}\overline{\Psi} = H\overline{\Psi}. \qquad (4.3.3)$$

# Elementary Applications

Equation 4.3.3 is identical to the time dependent Schrödinger equation. In fact, the above equation states that under time reversal, the system that was described by state $\Psi$ would become $\overline{\Psi}$. In this sense $\overline{\Psi}$ is the time reversed state of the system. Under time reversal, various physical quantities related to the system may behave differently (as a simple example, for a particle moving in space-time, under time reversal, assuming there are no spatial transformations involved, the coordinates of the particle remain the same no matter how its velocity vector reverses in direction). The physical quantity $\mathcal{O}$ itself may remain invariant or change sign upon time reversal. This behaviour is important in determining the selection rules for transitions among states that are diagonal with respect to energy. Let the closed system under consideration be in a stationary state whose basis states transform according to the irreducible representation $\Gamma$. If the degree of the representation $\Gamma$ is $d$, then the energy level is $d$-fold degenerate with $\{\psi_1, \cdots, \psi_d\}$ serving as a basis. It is of course assumed that all the basis states are real. If $\psi$ is the state of the system belonging to irreducible representation $\Gamma$, then in general $\psi$ is a linear combination of the basis states:

$$\psi = \sum_{m=1}^{d} c_m \psi_m.$$

The average value of the quantity $\mathcal{O}$ in this state is obtained according to the usual rules of quantum mechanics. If $\langle \mathcal{O} \rangle$ represents the average value then

$$\langle \mathcal{O} \rangle = \int \overline{\psi} \mathcal{O} \psi = \sum_{m,n=1}^{d} \overline{c_m} c_n \mathcal{O}_{mn}.$$

In the time reversed state $\overline{\psi}$, the average value of $\mathcal{O}$ would be obtained to be

$$\langle \mathcal{O} \rangle_{\text{rev}} = \int \overline{\overline{\psi}} \mathcal{O} \overline{\psi} = \sum_{m,n=1}^{d} \overline{c_m} c_n \mathcal{O}_{nm}.$$

If the physical quantity represented by the operator $\mathcal{O}$ was such that it remained invariant under time reversal, then its average value would remain unchanged under such a transformation. The value $\langle \mathcal{O} \rangle$ and $\langle \mathcal{O} \rangle_{\text{rev}}$ would then be equal. As the coefficients $c_i$ can be complex, it follows that

$$\mathcal{O}_{mn} = \mathcal{O}_{nm}$$

$$\Rightarrow \int \psi_m \mathcal{O} \psi_n = \int \psi_n \mathcal{O} \psi_m.$$

The transition matrix elements would therefore be non-zero when $\Gamma^{\mathcal{O}}$ is present in $(\Gamma \otimes \Gamma)^{\sigma}$, the symmetric component of the tensor product. For a physical quantity that changes its sign under time reversal, the matrix elements would be non-zero if the antisymmetric component $(\Gamma \otimes \Gamma)^\alpha$ contained $\Gamma^O$. This detour on time reversal action leading to the complex conjugation of wavefunction is generally known as *antilinear* property within the linear algebra context.

**Example 32.** Consider the character table for $C_{4v}$.

| $C_{4v}$ | $E$ | $2C_4$ | $C_2$ | $2\sigma_v$ | $2\sigma_d$ | | |
|---|---|---|---|---|---|---|---|
| $A_1$ | 1 | 1 | 1 | 1 | 1 | $z$ | $x^2+y^2; z^2$ |
| $A_2$ | 1 | 1 | 1 | $-1$ | $-1$ | $R_z$ | |
| $B_1$ | 1 | $-1$ | 1 | 1 | $-1$ | | $x^2-y^2$ |
| $B_2$ | 1 | $-1$ | 1 | $-1$ | 1 | | $xy$ |
| $E$ | 2 | 0 | $-2$ | 0 | 0 | $(x,y); (R_x R_y)$ | $xz; yz$ |

The electric dipole moment **p** for a system of charges is defined to be $\sum q_i r_i$. For a single charge, the $x$-component of the dipole moment vector ($p_x$) clearly has transformation properties of the $x$ coordinate which belongs to the E representation of $C_{4v}$. Hence $\Gamma^{p_x} \in \Gamma^E$. Consider a transition from an initial state corresponding E representation to a final state corresponding to the $A_2$ representation due to $x$-component dipole moment interaction. The matrix element of $p_x$ corresponding to this transition would be non-zero only if the tensor product $A_2 \otimes E$ contained the representation E (the representation transforming $p_x$). It can be observed from the table given above that in fact $A_2 \otimes E = E$. Thus $\langle A_2 | q\hat{x} | E \rangle$ is non-zero and the dipole transition is possible. This selection rule of non-zero matrix element will also be valid for the $p_y$ operator because the $y$-component dipole moment belongs to the irreducible representation E as well. However, $p_z$ belongs to trivial representation and hence $\langle A_2 | q\hat{z} | E \rangle = 0$. □

**Example 33.** Consider the electric quadrupole moment tensor defined as

$$Q_{ik} = e(3x_i x_k - r^2 \delta_{ik}), \tag{4.3.4}$$

where the indices $i$ and $k$ take values 1, 2 and 3. In the coordinate notation used in the text, $x_1 = x$, $x_2 = y$, $x_3 = z$ and $r^2 = x^2 + y^2 + z^2$. The tensor is clearly symmetric. Explicitly, the nine components of the quadrupole tensor for unit charge $e$ are

$$Q_{xx} \sim 3x^2 - r^2, \quad Q_{yy} \sim 3y^2 - r^2, \quad Q_{zz} \sim 3z^2 - r^2,$$

$$Q_{xx} + Q_{yy} + Q_{zz} = 0,$$

$$(Q_{xy} = Q_{yx}) \sim 3xy, \quad (Q_{xz} = Q_{zx}) \sim 3xz, \quad (Q_{yz} = Q_{zy}) \sim 3yz.$$

From above relationships, it follows that $Q_{ik}$ has 5 independent quantities. It is our purpose to study quadrupole moment selection rules in a system having the octahedral ($O$) symmetry. The character table for $O$ symmetry group follows:

| $O$ | $E$ | $8C_3$ | $3C_2$ | $6C_4$ | $6C_2$ | | |
|---|---|---|---|---|---|---|---|
| $A_1$ | 1 | 1 | 1 | 1 | 1 | | $x^2+y^2+z^2$ |
| $A_2$ | 1 | 1 | 1 | $-1$ | $-1$ | | |
| $E$ | 2 | $-1$ | 2 | 0 | 0 | 0 | $(2z^2-x^2-y^2, x^2-y^2)$ |
| $T_1$ | 3 | 0 | $-1$ | 1 | $-1$ | $(x,y,z)$ $(R_x, R_y, R_z)$ | |
| $T_2$ | 3 | 0 | $-1$ | $-1$ | 1 | | $(xy, xz, yz)$ |

Comparing the bases of the representation $T_2$, it is clear that the triplet $(Q_{xy}, Q_{xz}, Q_{yz})$ transforms according to the $T_2$ representation of $O$. Also notice that $Q_{zz} \sim 2z^2 - x^2 - y^2$ and $Q_{xx} - Q_{yy} \sim x^2 - y^2$. Therefore pair $(Q_{zz}, Q_{xx} - Q_{yy})$ transforms according to the E representation of $O$.

Let us first find which matrix elements of $(Q_{xy}, Q_{xz}, Q_{yz})$ are non zero for transitions between states of different representations. For this purpose one needs to find which irreducible representations are present in the tensor product of $T_2$ representation with various other irreducible representations of the group. This is accomplished, as has already been shown through previous examples, by use of Equation 3.7.2. We will leave it to the readers to verify the following:

$T_2 \otimes A_1 = T_2,$

$T_2 \otimes A_2 = T_1,$

$T_2 \otimes E = T_1 \oplus T_2,$

$T_2 \otimes T_1 = A_2 \oplus E \oplus T_1 \oplus T_2,$

$T_2 \otimes T_2 = A_1 \oplus E \oplus T_1 \oplus T_2.$

From the above decomposition, it is obvious that $Q_{xy}, Q_{xz}, Q_{yz}$ have non-zero matrix elements for transitions

$A_1 \leftrightarrow T_2; A_2 \leftrightarrow T_1; E \leftrightarrow T_1, T_2; T_1 \leftrightarrow T_2.$

To calculate the non-zero matrix elements for $Q_{xx}, Q_{yy}, Q_{zz}$, proceeding as above, the following decompositions of the tensor product of the E representation with various other representations of the group are obtained.

$E \otimes A_1 = E,$

$E \otimes A_2 = E,$

$E \otimes E = A_1 \oplus A_2 \oplus E,$

$$E \otimes T_1 = T_1 \oplus T_2,$$

$$E \otimes T_2 = T_1 \oplus T_2.$$

Once again it follows from the above decomposition that $Q_{xx}$, $Q_{yy}$, $Q_{zz}$ have non zero matrix elements for transitions

$$E \leftrightarrow A_1, A_2; T_1 \leftrightarrow T_2.$$

Till now, the transition matrix elements between states of different representations have been considered. In order to consider matrix elements that are diagonal with respect to energy, the first thing to note is that electric quadrupole moment tensor remains invariant under time reversal. This follows directly from the form of $Q_{ik}$ which is itself defined in terms of quantities that remain unaffected under time reversal. Therefore the symmetric component of the tensor product of an irreducible representation with itself needs to be considered. As an example, consider the symmetric component of the tensor product $T_2 \otimes T_2$, and let this component be represented $(T_2 \otimes T_2)^\sigma$ (in the notation of Chapter 3, Section 3.7). The class characters in the representation $(T_2 \otimes T_2)^\sigma$ are easily found using Equation 3.7.5.

$$\chi_\sigma^2(E) = 6,\ \chi_\sigma^2(8C_3) = 0,\ \chi_\sigma^2(3C_2) = 2,\ \chi_\sigma^2(6C_4) = 0,\ \chi_\sigma^2(6C_2) = 2.$$

From the above character values, the decomposition of $(T_2 \otimes T_2)^\sigma$ is obtained to be

$$(T_2 \otimes T_2)^\sigma = A_1 \oplus E \oplus T_2.$$

In a similar manner, the symmetric components of the tensor products of the other irreducible representations may be obtained.

$$(A_1 \otimes A_1)^\sigma = A_1,$$

$$(A_2 \otimes A_2)^\sigma = A_1,$$

$$(E \otimes E)^\sigma = A_1 \oplus E,$$

$$(T_1 \otimes T_1)^\sigma = A_1 \oplus E \oplus T_2,$$

$$(T_2 \otimes T_2)^\sigma = A_1 \oplus E \oplus T_2.$$

The matrix elements of $Q_{xx}$, $Q_{yy}$, $Q_{zz}$ that are diagonal with respect to energy are non-zero for transitions in E, $T_1$ and $T_2$. Similarly, the matrix elements of $Q_{xy}$, $Q_{xz}$, $Q_{yz}$ that are diagonal with respect to energy are non-zero for transitions in $T_1$ and $T_2$. □

The above two examples indicate the power of group theory concepts in deducing selection rules. We have clearly demonstrated that the group representation theory can determine whether any matrix elements of self-adjoint operators between stationary states is zero or non-zero. In the following section, we will exploit group theory tools to determine the independent vibrational modes of molecules possessing discrete group symmetry. These are well known as normal modes which characterize the molecules.

## 4.4 Molecular Vibrations

We will first briefly review the concepts of normal modes of vibrations of simple systems which are well known in classical mechanics. Then we will elaborate the elegant group theory approach in reproducing those normal modes of molecules.

Consider the classical problem of two masses $m_1$ and $m_2$ connected by a Hooke's spring of stiffness $\varkappa$. It is well known that a small disturbance/displacement leads to oscillations/vibration of the system with a unique frequency $\omega$ given by

$$\omega = \sqrt{\frac{\varkappa}{\mu}}.$$

In the above expression for $\omega$, $\mu\left(=\frac{m_1 m_2}{m_1+m_2}\right)$ is the reduced mass of the system. When the number of participating masses and springs is increased, the number of degrees of freedom also increases and as a result the oscillatory motion of the system becomes more complicated. Consider a system consisting of $n$ mass points. Such a system of mass points has in general $3n$ degrees of freedom of movement. Of these, 3 degrees of freedom correspond to an overall motion of the system in the three space directions. Additionally, the system will have rotational degrees of freedom. If the distribution of the mass points in the equilibrium state is in three-dimensional space, then the system has 3 rotational degrees of freedom. All the other degrees correspond to the vibrational motion and therefore such a system has a total of $3n - 6$ vibrational degrees of freedom. In case the mass points were distributed along a straight line, the rotational degree of freedom associated with rotation about the axis of the system would have to be discounted since here it is assumed that the masses have negligible physical dimensions. For such an arrangement of mass points, there will be then a total of $3n - 5$ vibrational degrees of freedom.

Let us recall the basic theory of small oscillations from classical mechanics. Consider a system with $s$ vibrational degrees of freedom. Let $x_i (i = 1, 2, \cdots, s)$ denote small excursions of the mass points from their mean positions. Let x be a $s \times 1$ column vector such that $\{x\}_{i1} = x_i$. In general, the kinetic energy $T$ of the system is a quadratic function of $\dot{x}_i$ (time derivative of $x_i$). The most general form of $T$ is then given by

$$T = \frac{1}{2}\sum_{j,k}^{s} m_{jk}\dot{x}_j \dot{x}_k,$$

where $m_{jk}$ are symmetric constants ($m_{jk} = m_{kj}$). It is expected that the system will exhibit oscillatory motion when disturbed from its mean position. In this circumstance, the potential energy of the system can be taken to be minimum (and equal to 0) in the undisturbed state. When the excursions $x_i$ are small, then the potential energy $U$ may be written as,

$$U = \frac{1}{2} \sum_{j,k}^{s} \varkappa_{jk} x_j x_k,$$

where $\varkappa_{jk} (= (\frac{\partial^2}{\partial x_j \partial x_k} U)_{x_j=0, x_k=0})$ are some other symmetric constants. Let M and K be symmetric matrices such that $\{M\}_{jk} = m_{jk}$ and $\{K\}_{jk} = \varkappa_{jk}$. The Lagrangian $L$ of the system can therefore be written as

$$L = T - U = \frac{1}{2} \sum_{j,k}^{s} m_{jk} \dot{x}_j \dot{x}_k - \frac{1}{2} \sum_{j,k}^{s} \varkappa_{jk} x_j x_k$$

$$\Rightarrow L = \frac{1}{2} \left[ \dot{x}^T M \dot{x} - x^T K x \right]. \tag{4.4.1}$$

We note the Euler–Lagrange equation for convenience here.

$$\frac{d}{dt} \left( \frac{\partial}{\partial \dot{x}_i} L \right) = \frac{\partial}{\partial x_i} L.$$

The equations of motion are obtained by substituting the form of $L$ in the Euler–Lagrange equation. It may be noted that

$$\frac{\partial}{\partial \dot{x}_i} L = \frac{1}{2} \sum_{k=1}^{s} (m_{ik} + m_{ki}) \dot{x}_k = \sum_{k=1}^{s} m_{ik} \dot{x}_k,$$

because of symmetry of $m_{ik}$. In a similar manner $\frac{\partial}{\partial x_i} L$ can be obtained. The equations of motion in their final form are

$$\sum_{k=1}^{s} (m_{ik} \ddot{x}_k + \varkappa_{ik} x_k) = 0; \quad i = (1, 2 \cdots, s). \tag{4.4.2}$$

Equation 4.4.2 admits simple harmonic solutions of the form $x_k = a_k \exp(i\omega t)$. Substituting for $x_k$ in the above system of equations, we obtain a homogeneous system of equations in complex quantities $a_k$.

$$\sum_{k=1}^{s} \left( -m_{ik} \omega^2 + \varkappa_{ik} \right) a_k = 0; \quad i = (1, 2 \cdots, s). \tag{4.4.3}$$

The above system has non-trivial solutions for complex quantities $a_k$ if the determinant of the coefficient matrix vanishes.

$$\begin{vmatrix} \varkappa_{11} - \omega^2 m_{11} & \cdots & \varkappa_{1n} - \omega^2 m_{1n} \\ \vdots & \ddots & \vdots \\ \varkappa_{n1} - \omega^2 m_{n1} & \cdots & \varkappa_{nn} - \omega^2 m_{nn} \end{vmatrix} = 0. \tag{4.4.4}$$

Equation 4.4.4 is called the *secular equation* while the determinant in the left hand side is called the *secular determinant*. Upon explicitly expanding the secular determinant, a polynomial of degree $n$ in $\omega^2$ is obtained. All the values of $\omega^2$ obtained from solving this polynomial equation can be shown to be positive, but we do not prove this here. The positive roots of the various values of $\omega^2$ are called the *eigenfrequencies* of the system. For a system with $s$ vibrational degrees of motion, there are $s$ possible eigenfrequencies. It may so happen that some of the eigenfrequencies coincide, in which case they are called *degenerate*. Suppose that the system has no degenerate frequencies. Upon substituting each of the values of $\omega$, say $\omega_l$ in Equation 4.4.3, the values for $a_k$ are obtained up to the same constant complex multiplier $c_l$ for all $a_k$. Let this value of $a_k$ be denoted by $a_{kl}c_l$ so that $x_k^{(l)} = a_{kl}c_l \exp(i\omega_l t)$ where $a_{kl}$ are all real. Noting that Equation 4.4.2 is homogeneous and linear, the general solution may now be obtained as a linear combination of solutions corresponding to each eigenfrequency.

$$x_k = \Re \sum_{l=1}^{s} a_{kl}[c_l \exp(i\omega_l t)], \qquad (4.4.5)$$

where $\Re$ indicates the real part. With the definition $\eta_l = \Re c_l \exp(i\omega_l t)$, and representing the excursion $x_k$'s and $\eta_l$'s as $s \times 1$ column vectors x and $\eta$ respectively, the above equation may be written as a matrix equation

$$x = A\eta, \qquad (4.4.6)$$

where A is the coefficient matrix such that $[A]_{kl} = a_{kl}$. As the $x_k$'s are independent and so are $\eta_l$'s, the above equation can be inverted to express $\eta_l$ in terms of $x_k$'s. This allows us to choose a different set of generalized coordinates, namely $\eta_l$ to describe the motion of the system. The important property which makes this choice of generalized coordinates more convenient is that

$$\ddot{\eta}_l + \omega_l^2 \eta_l = 0; \quad l = (1, 2, \cdots, s).$$

Thus the description of the motion of the system is much simpler in terms of the coordinates $\eta_l$. These coordinates are called the *normal coordinates*. A normal coordinate is essentially a linear combination of the original coordinates that oscillates at a unique eigenfrequency. The motion of the system corresponding to a normal coordinate is called a *normal mode*. It may be noted that if some particular eigenfrequency $\omega_l$ was degenerate with a multiplicity $\nu_l$, then there would be $\nu_l$ normal coordinates associated with that frequency. More generally

$$\ddot{\eta}_{lp} + \omega_l^2 \eta_{lp} = 0, \quad \sum_l \nu_l = s, \qquad (4.4.7)$$

where the index $l$ ranges over the number of distinct eigen frequencies and the index $p$ ranges from 1 to $\nu_l$. Note that the range of $p$ varies with $l$. Applying the transformation in Equation 4.4.6, it is now possible to express the Lagrangian and the total energy $H$ of the vibrating system in a matrix form.

$$L = \frac{1}{2}\left[\dot{\eta}^T \left(A^T M A\right) \dot{\eta} - \eta^T \left(A^T K A\right) \eta\right], \tag{4.4.8}$$

$$H = \frac{1}{2}\left[\dot{\eta}^T \left(A^T M A\right) \dot{\eta} + \eta^T \left(A^T K A\right) \eta\right].$$

The important consequence of the transformation to normal coordinates is that the symmetric matrices M and K are simultaneously diagonalized, whereby leading to an expression of the energy of the vibrating system as a sum of energies associated with each of the normal vibrational modes. As a result, one has

$$H = \frac{1}{2}\left[\sum_l \sum_{p=1}^{v_l} \mathfrak{m}_{l,p}\dot{\eta}_{l,p}^2 + \sum_l \omega_l^2 \sum_{p=1}^{v_l} \mathfrak{m}_{l,p}\eta_{l,p}^2\right],$$

where $\mathfrak{m}_{l,p}$ are positive constants. It is conventional to redefine the normal coordinates as $\Theta_{l,p} = \sqrt{\mathfrak{m}_{l,p}}\eta_{l,p}$ so that the total energy can be written simply as

$$H = \frac{1}{2}\left[\sum_l \sum_{p=1}^{v_l} \dot{\Theta}_{l,p}^2 + \sum_l \omega_l^2 \sum_{p=1}^{v_l} \Theta_{l,p}^2\right]. \tag{4.4.9}$$

We will apply the classical mechanics approach on a triatomic molecule example and obtain the normal modes.

**Example 34.** (Landau and Lifshitz, *Mechanics*) Consider a non-linear triatomic molecule such as the one shown in Figure 4.4.1(a). The three mass point $m$, $M$ and $m$ are nothing but the nuclei of the atoms constituting the molecule. The size of the nuclei is negligible compared to the size of the molecule itself. As the three mass points will always lie in a plane, we need to consider motions that occur only in the *x-z* plane. For motions restricted to the plane, there are a total of 6 degrees of freedom, out of which 2 are translational and 1 rotational (about an axis perpendicular to the plane of motion). This leaves a total of $6 - 3 = 3$ vibrational degrees of freedom. Let the displacements of the atoms of mass $m$ from their mean positions be $(x_1, z_1)$ and $(x_2, z_2)$ and that of $M$ be $(X, Z)$. The overall translation of the molecule may be eliminated by imposing the condition that the center of mass of the molecule does not get displaced.

$$m(x_1 + x_2) + MX = 0,$$

$$m(z_1 + z_2) + MZ = 0. \tag{4.4.10}$$

As is obvious from the Figure 4.4.1(a), the location of the center of mass is $(0, \frac{2ml\cos\alpha}{2m+M})$ where $l$ is the bond length between $M$ and $m$ in the undisturbed state and $\alpha$ is the bond angle. The rotational motion of the molecule may be eliminated by equating the angular momentum of the molecule (with respect to the center of mass of the molecule) to zero. In case of small oscillations, the instantaneous positions of the atoms from the center of mass can always be approximated by their positions with respect to the center of mass in the undisturbed state.

$$\left(-\frac{2ml\cos\alpha}{2m+M}\hat{k}\right) \times M\left(\dot{Z}\hat{k} + \dot{X}\hat{i}\right) +$$

$$\left(l\sin\alpha\hat{i} + \frac{Ml\cos\alpha}{2m+M}\hat{k}\right) \times m\left(\dot{x}_1\hat{i} + \dot{z}_1\hat{k}\right) +$$

$$\left(-l\sin\alpha\hat{i} + \frac{Ml\cos\alpha}{2m+M}\hat{k}\right) \times m\left(\dot{z}_2\hat{k} + \dot{x}_2\hat{i}\right) = 0.$$

Simplifying the above by use of the first of Equation 4.4.10, the angular motion is eliminated completely by the condition

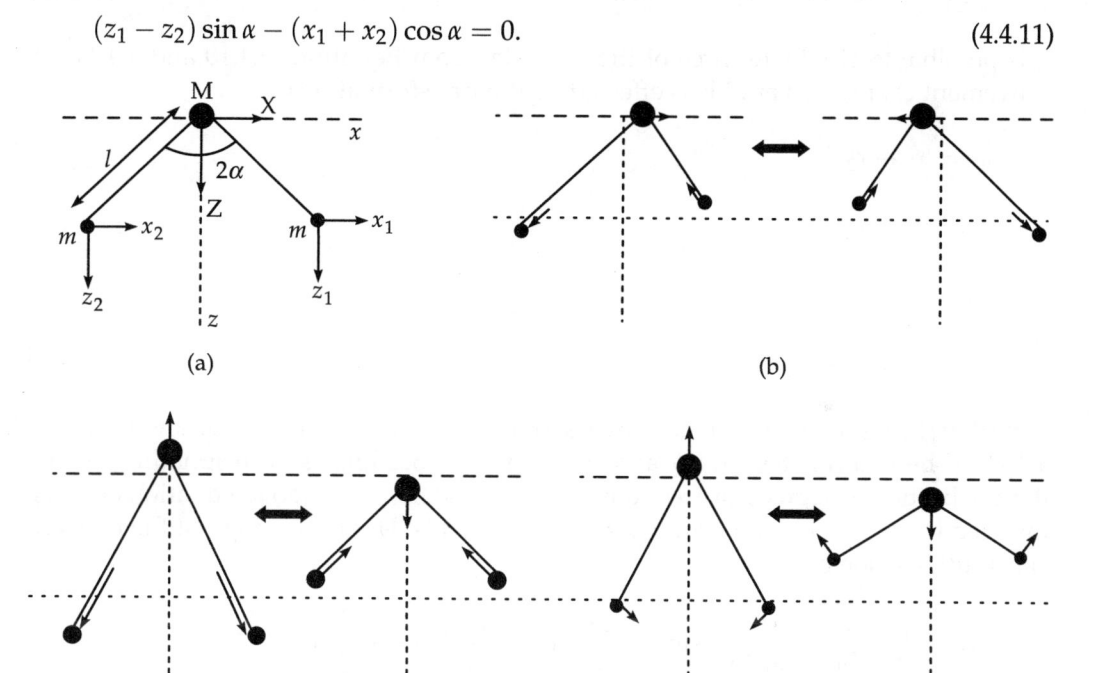

$$(z_1 - z_2)\sin\alpha - (x_1 + x_2)\cos\alpha = 0. \qquad (4.4.11)$$

**Figure 4.4.1**  Normal Modes (Non-linear Tri-atomic Molecule)

As long as displacements of the masses satisfy Equations 4.4.10 and 4.4.11, the motion will be purely vibrational. As the molecule vibrates, it will generally get deformed. There are essentially two types of deformations possible in the current case:

(1) The bond length $l$ changes for either or both the bonds
(2) The bond angle $2\alpha$ changes.

With both such deformations, we can associate springs of stiffness $\varkappa_l$ and $\varkappa_\alpha$. It is easily seen that the associated deformations are

$$\delta l_1 = (x_1 - X) \sin \alpha + (z_1 - Z) \cos \alpha,$$

$$\delta l_2 = (X - x_2) \sin \alpha + (z_2 - Z) \cos \alpha, \tag{4.4.12}$$

by considering the components of the displacements of atoms along the bond lengths. Likewise, considering the displacements perpendicular to the bond length, the change in the bond angle is obtained to be

$$l\delta(2\alpha) = (x_1 - X) \cos \alpha + (Z - z_1) \sin \alpha$$

$$- (x_2 - X) \cos \alpha - (z_2 - Z) \sin \alpha. \tag{4.4.13}$$

It is possible to eliminate three of the variables from Equations 4.4.10 and 4.4.11. A convenient change of variable is effected by the transformations

$$q_1 = x_1 + x_2,$$

$$q_2 = x_1 - x_2,$$

$$q_3 = z_1 + z_2,$$

$$q_1 \cot \alpha = z_1 - z_2. \tag{4.4.14}$$

Note that the last equation in the above is a consequence of the first three and Equation 4.4.11. The coordinates $q_1$, $q_2$ and $q_3$ are three coordinates corresponding to the three vibrational degrees of freedom of the molecule. The above equations can be inverted to obtain $z_1$, $z_2$, $Z$, $x_1$, $x_2$, $X$ in terms of $q_i$'s. Upon carrying out the relevant substitutions, one has

$$\delta l_1 = \frac{q_1}{2}\left[\frac{2m}{M} + \frac{1}{\sin^2\alpha}\right]\sin\alpha + \frac{q_2}{2}\sin\alpha + \frac{q_3}{2}\left[1 + \frac{2m}{M}\right]\cos\alpha,$$

$$\delta l_2 = -\frac{q_1}{2}\left[\frac{2m}{M} + \frac{1}{\sin^2\alpha}\right]\sin\alpha + \frac{q_2}{2}\sin\alpha + \frac{q_3}{2}\left[1 + \frac{2m}{M}\right]\cos\alpha,$$

$$l\delta(2\alpha) = q_2 \cos\alpha - q_3\left[1 + \frac{2m}{M}\right]\sin\alpha. \tag{4.4.15}$$

## Elementary Applications

The potential energy $U$ of the deformed molecule is calculated from $U = \frac{\varkappa_l}{2}[(\delta l_1)^2 + (\delta l_2)^2] + \frac{\varkappa_\alpha (l\delta(2\alpha))^2}{2}$. The reader is advised to carry out rather straightforward calculations leading to the following form of the Lagrangian.

$$L = \frac{m}{4}\left[\frac{2m}{M} + \frac{1}{\sin^2 \alpha}\right]\dot{q}_1^2 - \frac{\varkappa_l}{4}\left[\frac{2m}{M} + \frac{1}{\sin^2 \alpha}\right]_2^2 q_1^2 \sin^2 \alpha +$$

$$\frac{m}{4}\dot{q}_2^2 - \left[\varkappa_l \sin^2 \alpha + 2\varkappa_\alpha \cos^2 \alpha\right]\frac{q_2^2}{4} +$$

$$\frac{m}{4}\left[1 + \frac{2m}{M}\right]\dot{q}_3^2 - \left[1 + \frac{2m}{M}\right]^2 \left[\varkappa_l \cos^2 \alpha + 2\varkappa_\alpha \sin^2 \alpha\right]\frac{q_3^2}{4} +$$

$$\left[1 + \frac{2m}{M}\right][\varkappa_l - 2\varkappa_\alpha]\frac{q_2 q_3}{2} \sin \alpha \cos \alpha. \tag{4.4.16}$$

Finally, the equations of motion obtained from the above form of the Lagrangian are

$$m\ddot{q}_1 + \left[1 + \frac{2m}{M}\sin^2 \alpha\right]\varkappa_l q_1 = 0,$$

$$m\ddot{q}_2 + \left[\varkappa_l \sin^2 \alpha + 2\varkappa_\alpha \cos^2 \alpha\right]q_2 +$$

$$+ \left[1 + \frac{2m}{M}\right][\varkappa_l - 2\varkappa_\alpha](\sin \alpha \cos \alpha)q_3 = 0,$$

$$m\ddot{q}_3 + \left[1 + \frac{2m}{M}\right]\left[\varkappa_l \cos^2 \alpha + 2\varkappa_\alpha \sin^2 \alpha\right]q_3 +$$

$$+ [\varkappa_l - 2\varkappa_\alpha](\sin \alpha \cos \alpha)q_2 = 0. \tag{4.4.17}$$

It is clear that the coordinate $q_1 = x_1 + x_2$ oscillates normally with a frequency $\sqrt{\frac{\varkappa_l}{m}[1 + \frac{2m}{M}\sin^2 \alpha]}$. When only this particular mode is excited in the molecule, then $q_2 = q_3 = 0$, or in other words $x_1 = x_2$ and $z_1 = -z_2$. During such an oscillation, the central atom of mass $M$ does not get displaced in the z-direction. This mode is depicted in Figure 4.4.1 (b).

The other coordinates $q_2$ and $q_3$ are coupled. The secular equation for the eigenfrequencies of the normal modes associated with $q_2$ and $q_3$ is

$$\begin{vmatrix} [\varkappa_l \sin^2 \alpha + 2\varkappa_\alpha \cos^2 \alpha] - \omega^2 m & [1 + \frac{2m}{M}][\varkappa_l - 2\varkappa_\alpha] \times (\sin \alpha \cos \alpha) \\ [\varkappa_l - 2\varkappa_\alpha] \times (\sin \alpha \cos \alpha)v_2 & [1 + \frac{2m}{M}][\varkappa_l \cos^2 \alpha + 2\varkappa_\alpha \sin^2 \alpha] - \omega^2 m \end{vmatrix} = 0.$$

The solution of the secular equation gives 2 more eigenfrequencies corresponding to distinct normal modes. These modes would in general be some linear combinations of $q_2$ and $q_3$. When only one of these modes is excited then $q_1 = 0$, i.e., $x_1 = -x_2$. The central atom of mass $M$ cannot get displaced in the $x$-direction now. Figure 4.4.1 (c), (d) depicts the two modes. □

It will be interesting to obtain all the three non-degenerate normal modes depicted in Figure 4.4.1 (b), (c), (d) using group theory. Clearly, this molecule has $C_{2v}$ symmetry. We revisit this triatomic molecule (see Example 36) where we reproduce these normal modes by exploiting the group symmetry. We will elaborate another example using the conventional classical mechanics procedure to emphasize that the methodology is tedious but systematic. Then, the readers will be able to appreciate the results when we discuss elegant group theory approaches towards deriving normal modes.

**Example 35.** Consider a hypothetical equilateral molecule of three identical atoms shown in Figure 4.4.2. The Lagrangian of the system is given by

$$L = \frac{m}{2}\left[\dot{x}_1^2 + \dot{y}_1^2 + \dot{x}_2^2 + \dot{y}_2^2 + \dot{x}_3^2 + \dot{y}_3^2\right] - \frac{\varkappa}{2}\left[\Delta l_{12}^2 + \Delta l_{23}^2 + \Delta l_{13}^2\right],$$

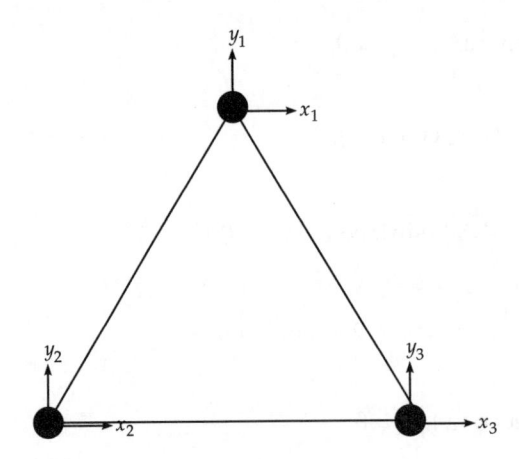

**Figure 4.4.2** Hypothetical Equilateral Molecule

where $\Delta l_{12}$ is the extension in the bond between nuclei 1 and 2 etc. The extensions are easily calculated to be

$$\Delta l_{12} = \frac{x_1 - x_2}{2} + \sqrt{3}\frac{y_1 - y_2}{2},$$

$$\Delta l_{13} = \frac{x_3 - x_1}{2} + \sqrt{3}\frac{y_1 - y_3}{2},$$

$$\Delta l_{23} = x_3 - x_2.$$

# Elementary Applications

The conditions for zero momentum of the molecule and zero angular momentum are given by

$$x_1 + x_2 + x_3 = 0,$$

$$y_1 + y_2 + y_3 = 0,$$

$$-x_1 + \frac{x_2 + x_3}{2} + \sqrt{3}\frac{y_3 - y_2}{2} = 0.$$

With a choice of $q_1 = x_1$, $q_2 = x_3 - x_2$ and $q_3 = y_1 - y_3$ and from above equations, three coordinates can be eliminated. From here the Lagrangian can be expressed completely in terms of the three vibrational coordinates $q_1$, $q_2$, $q_3$ and their time derivatives. The intervening steps are left as an exercise for the reader who shall eventually obtain the following equations of motion:

$$\ddot{q}_1 + \frac{3\varkappa}{2m}q_1 = 0,$$

$$\frac{3\varkappa}{4m}q_1 + \left(\ddot{q}_2 + \frac{9\varkappa}{4m}q_2\right) + \frac{\sqrt{3}\varkappa}{2m}q_3 = 0,$$

$$\frac{3\sqrt{3}\varkappa}{8m}q_1 + \frac{3\sqrt{3}\varkappa}{8m}q_2 + \left(\ddot{q}_3 + \frac{9\varkappa}{4m}q_3\right) = 0.$$

The secular equation of this system gives a doubly degenerate mode of frequency $\sqrt{\frac{3\varkappa}{2m}}$ and another mode of frequency $\sqrt{\frac{3\varkappa}{m}}$. □

The technique illustrated in Examples 34 and 35 for calculating eigenfrequencies of molecular vibrations is generalizable for molecules with more atoms. A thorough calculation as performed in the previous examples is desirable for all these polyatomic molecules. However, the computation becomes very tedious. In many situations, it is not the exact values of these eigenfrequencies that are needed but rather a classification of the normal modes. In the case of the non-linear triatomic molecule, this classification could have been done without actually going through the full calculations. With reference to Figure 4.4.1, one may note that the three normal modes of the motion are quite obvious from mere inspection of the structure of the molecule. The symmetry of the molecule allows one to guess the simplest possible ways in which the molecule can be vibrating. While such intuitive guesswork might work in the simplest of cases, a more formal approach is necessary when one needs to classify the normal modes of polyatomic molecules. Group theory provides an elegant solution to this classification problem.

In a vibrating molecule, the atomic nuclei do not remain in their mean positions. If the nuclei were assumed fixed in their mean positions, then a given molecule would possess a certain point group symmetry. For small vibrations of the molecule, the symmetry is amply retained so as to be useful in classification of the normal modes.

If the molecule is vibrating in some specific normal mode, upon a symmetry transformation the frequency of the oscillation should remain unchanged but the coordinate designations change for various nuclei. The vibration in the transformed state of the molecule can at best be a linear combination of the degenerate normal modes associated with this particular frequency. The degenerate normal coordinates of the molecule are transformed into linear combinations of each other upon action of the symmetry group and therefore they give a representation of the symmetry group. That this representation is always irreducible is ensured by the fact that the quadratic form $\sum_{p=1}^{v_l} \Theta_{l,p}^2$ (Equation 4.4.9) remains invariant under every symmetry transformation. For, if the representation of the symmetry group given by the normal coordinates of a degenerate frequency were reducible, then the cancellation of the cross terms in the transformed value of $\sum_{p=1}^{v_l} \Theta_{l,p}^2$ could no more be guaranteed. The conclusion may now be stated that the degenerate normal modes of vibration give an irreducible representation of the symmetry group of the molecule.

If the normal coordinates are known at the outset, it would be a simple matter to calculate the symmetry group action on them and thereby construct a representation of the same. Such a representation is called the *full vibrational representation* of the symmetry group of the molecule. The vibrational representation is a reducible representation. From the characters of the vibrational representation it is possible to decompose the representation into irreducible parts using a projection operator. Since the knowledge of the normal coordinates is not assumed to begin with, the characters of the vibrational representation would have to be calculated by other means. In Equation 4.4.6, it was noted that the transformation A is an invertible transformation. It follows that the representation of the symmetry group over the normal coordinates and the representation of the symmetry group over any other basis for the vibrational degrees of freedom are equivalent, and hence have the same characters. Once a basis for the vibrational degrees of freedom is obtained, the symmetry group action on the basis yields a representation of the symmetry group. The characters of the group elements can be easily calculated from the representing matrices.

**Example 36.** The non-linear triatomic molecule is considered here again. The symmetry group of the molecule is clearly $C_{2v}$. With reference to Figure 4.4.1(a), the $C_2$ axis is along the *z-axis*, the $\sigma_v(\sigma_v')$ planes are the *x-z* plane and the *y-z* plane. To each of the molecules is attached a basis of its excursions from the mean position, namely $(x_1, z_1)$, $(x_2, z_2)$ and $(X, Z)$. Upon eliminating the translational and rotational motion of the molecule it was shown in the previous example that coordinates $q_1 = x_1 + x_2$, $q_2 = x_1 - x_2$ and $q_3 = z_1 + z_2$ correspond to the three vibrational degrees of freedom, though not all of them are the normal coordinates. Consider the basis set $(q_1, q_2, q_3)$ and let the symmetry group $C_{2v}$ act on this basis. The representation of $C_{2v}$ on this basis is the full vibrational representation $\Gamma^V$. The reader should have no difficulty in verifying that the representing matrices for various group elements are

# Elementary Applications

$$\Gamma^V(E) = \begin{bmatrix} 1 & 0 & 0 \\ 0 & 1 & 0 \\ 0 & 0 & 1 \end{bmatrix}, \quad \Gamma^V(C_2) = \begin{bmatrix} -1 & 0 & 0 \\ 0 & 1 & 0 \\ 0 & 0 & 1 \end{bmatrix},$$

$$\Gamma^V(\sigma_v) = \begin{bmatrix} 1 & 0 & 0 \\ 0 & 1 & 0 \\ 0 & 0 & 1 \end{bmatrix}, \quad \Gamma^V(\sigma_v') = \begin{bmatrix} -1 & 0 & 0 \\ 0 & 1 & 0 \\ 0 & 0 & 1 \end{bmatrix}.$$

Note that the action of $C_2$ is such that $x_1 \mapsto -x_2, x_2 \mapsto -x_1, z_1 \mapsto z_1$ and $z_2 \mapsto z_2$; consequently $q_1 \mapsto -q_1, q_2 \mapsto q_2$ and $q_3 \mapsto q_3$. Likewise the whole group action can be worked out. The character table for $C_{2v}$ that includes the characters for the representation $\Gamma^V$ is

| $C_{2v}$ | $E$ | $C_2$ | $\sigma_v$ | $\sigma_v'$ | |
|---|---|---|---|---|---|
| $A_1$ | 1 | 1 | 1 | 1 | $z$ |
| $A_2$ | 1 | 1 | $-1$ | $-1$ | $R_z$ |
| $B_1$ | 1 | $-1$ | 1 | $-1$ | $x, R_y$ |
| $B_2$ | 1 | $-1$ | $-1$ | 1 | $y, R_x$ |
| $\Gamma^V$ | 3 | 1 | 3 | 1 | $(q_1, q_2, q_3)$ |

From the character table, it is easily seen that $\Gamma^V = A_1 \oplus A_1 \oplus B_1$. Since all the irreducible representations present in $\Gamma^V$ are of degree 1, all the eigenfrequencies are non-degenerate. Two of the eigenfrequencies transform according to the symmetric representation $A_1$, the ones depicted in Figures 4.4.1(c) and (d). Then there is one frequency corresponding to the antisymmetrical representation $B_1$ of Figure 4.4.1(b). □

Instead of generating the vibrational representation of the molecule, it is often more convenient to generate the representation on a basis consisting of excursions. We do this in the following example.

**Example 37.** Continuing from the previous example, let the excursions be labeled $(x_1, z_1)$, $(x_2, z_2)$ and $(X, Z)$. Consider the group action on the basis $(x_1, z_1, x_2, z_2, X, Z)$. In this case, identify the degree 6 reducible representation as $\Gamma^M$. The following representing matrices are obtained for $\Gamma^M$:

$$\Gamma^M(E) = \begin{bmatrix} 1 & 0 & 0 & 0 & 0 & 0 \\ 0 & 1 & 0 & 0 & 0 & 0 \\ 0 & 0 & 1 & 0 & 0 & 0 \\ 0 & 0 & 0 & 1 & 0 & 0 \\ 0 & 0 & 0 & 0 & 1 & 0 \\ 0 & 0 & 0 & 0 & 0 & 1 \end{bmatrix},$$

$$\Gamma^M(C_2) = \begin{bmatrix} 0 & 0 & -1 & 0 & 0 & 0 \\ 0 & 0 & 0 & 1 & 0 & 0 \\ -1 & 0 & 0 & 0 & 0 & 0 \\ 0 & 1 & 0 & 0 & 0 & 0 \\ 0 & 0 & 0 & 0 & -1 & 0 \\ 0 & 0 & 0 & 0 & 0 & 1 \end{bmatrix},$$

$$\Gamma^M(\sigma_v) = \begin{bmatrix} 1 & 0 & 0 & 0 & 0 & 0 \\ 0 & 1 & 0 & 0 & 0 & 0 \\ 0 & 0 & 1 & 0 & 0 & 0 \\ 0 & 0 & 0 & 1 & 0 & 0 \\ 0 & 0 & 0 & 0 & 1 & 0 \\ 0 & 0 & 0 & 0 & 0 & 1 \end{bmatrix},$$

$$\Gamma^M(\sigma_v') = \begin{bmatrix} 0 & 0 & -1 & 0 & 0 & 0 \\ 0 & 0 & 0 & 1 & 0 & 0 \\ -1 & 0 & 0 & 0 & 0 & 0 \\ 0 & 1 & 0 & 0 & 0 & 0 \\ 0 & 0 & 0 & 0 & -1 & 0 \\ 0 & 0 & 0 & 0 & 0 & 1 \end{bmatrix}.$$

As for the characters of $\Gamma^M$, one has

| $C_{2v}$ | $E$ | $C_2$ | $\sigma_v$ | $\sigma_v'$ |
|---|---|---|---|---|
| $\Gamma^M$ | 6 | 0 | 6 | 0 |

Upon decomposition, it is found that $\Gamma^M = 3A_1 \oplus 3B_1$. Now consider the representations of pure translations and rotations that keep the molecule in the $x$-$z$ plane. Translations in $z$ and $x$ directions correspond to the representations $A_1$ and $B_1$ respectively, while the rotation about the $y$-axis corresponds to another instance of the irreducible representation $B_1$. The vibrational representation $\Gamma^V$ is therefore equal to $\Gamma^M - (A_1 \oplus 2B_1)$ which is same as $A_1 \oplus A_1 \oplus B_1$, as was found before for vibrational representation.

Thus having determined the irreducible representations representing the normal modes of the triatomic molecule possessing $C_{2v}$ symmetry, we can extract the three non-degenerate normal coordinates in the following manner: Construct the projection operators $P_{A_1}$ and subtract the translation operator $T_z$ denoting the translation along the $z$ direction which belongs to $A_1$ basis. The explicit matrix form of the two operators in the excursion basis is

$$\mathbf{P}^M_{A_1} = \frac{1}{4} = \begin{bmatrix} 2 & 0 & -2 & 0 & 0 & 0 \\ 0 & 2 & 0 & 2 & 0 & 0 \\ -2 & 0 & 2 & 0 & 0 & 0 \\ 0 & 2 & 0 & 2 & 0 & 0 \\ 0 & 0 & 0 & 0 & 0 & 0 \\ 0 & 0 & 0 & 0 & 0 & 4 \end{bmatrix}, \quad \mathbf{T}^M_z = \frac{1}{3} \begin{bmatrix} 0 & 0 & 0 & 0 & 0 & 0 \\ 0 & 1 & 0 & 1 & 0 & 1 \\ 0 & 0 & 0 & 0 & 0 & 0 \\ 0 & 1 & 0 & 1 & 0 & 1 \\ 0 & 0 & 0 & 0 & 0 & 0 \\ 0 & 1 & 0 & 1 & 0 & 1 \end{bmatrix}.$$

In fact, the rank of $\mathbf{P}_{A_1} - \mathbf{T}_z$ is two whose non-trivial eigenvectors

$$\begin{pmatrix} -1 \\ 0 \\ 1 \\ 0 \\ 0 \\ 0 \end{pmatrix} = x_2 - x_1, \quad \begin{pmatrix} 0 \\ -1/2 \\ 0 \\ -1/2 \\ 0 \\ 1 \end{pmatrix} = Z - \frac{z_1 + z_2}{2},$$

which are proportional to $q_2$ and $q_3$ in the center of the mass frame.

The projection operator eigenvectors are not in general the frequency eigenbasis. However, this group theory approach will assert that the frequency basis (normal modes) belonging to the $A_1$ irreducible representation will involve only linear superposition of $q_2$ and $q_3$. Similarly, we can obtain the projection operator for irreducible representation $P_{B_1}$ and subtract translation operator $T_x$. The rank of $P_{B_1} - T_x$ is again two whose the non-trivial eigenvectors are $q_1 = x_1 + x_2$ (vibrational mode) and $z_1 - z_2$ (rotational mode) consistent with the classical mechanics approach discussed in Example 34. □

The above two examples illustrate how the representation of the symmetry group of a molecule on the basis of excursions can be utilized to infer the normal modes of the vibrations of the molecule. Also, the matrix representation in the excursion basis is useful to infer the normal coordinates. However, the characters of the vibrational representation can be inferred directly in all cases without actually determining the representing matrices.

Consider a molecule with point group symmetry $G$ in a coordinate system oriented in the usual way so that $z$-axis is the symmetry axis of the molecule. With each nucleus of the molecule as an origin, construct coordinate systems of excursions such that their axes are parallel to the original coordinate system (Figure 4.4.1(a)). We consider the group action on the excursions. There are $3n$ coordinates in general for an $n$-atomic molecule. If a symmetry transformation displaces a nucleus into another identical one, then the representing matrix of the transformation cannot have diagonal elements corresponding to coordinates of the first nucleus. Thus the character of the transformation will depend on the action of the transformation on the coordinates of the nuclei which are unmoved from their positions by the transformation. Suppose the symmetry transformation $C_\theta$ is a rotation of the molecule about $z$-axis by an angle $\theta$.

Let the molecule whose excursion coordinates are $(x_a, y_a, z_a)$ remain unmoved by the $C_\theta$ transformation. The excursion coordinates are transformed due to rotation as

$$x_a \mapsto x_a \cos\theta + y_a \sin\theta,$$
$$y_a \mapsto -x_a \sin\theta + y_a \cos\theta,$$
$$z_a \to z_a.$$

The contribution to the character of the rotation $C_\theta$ from one such nucleus is $(1 + 2\cos\theta)$ If there are $N_C$ nuclei which are unmoved, then the total contribution to the character is $N_C(1 + 2\cos\theta)$. However, in the process of calculating the character, one has to keep track of the translational motion of the molecule as a whole as well as its rotation about the symmetry axis. Under pure rotation, both the vectors corresponding to small displacements of the center of mass as well as small angular displacements behave as polar vectors and consequently contribute $2(1 + 2\cos\theta)$ to the character of $C_\theta$. Therefore the character of $C_\theta$ in the vibrational representation $\Gamma^V$ is given by

$$\chi^{\Gamma^V}(C_\theta) = (N_C - 2)(1 + 2\cos\theta). \tag{4.4.18}$$

For calculating the character of the rotary reflection transformation $S_\theta$, it may be noted first of all that under such a transformation the coordinates $x_a$ and $y_a$ transform as indicated above, however $z_a \mapsto -z_a$. If there are $N_S$ nuclei which are unmoved by $S_\theta$, then the total contribution to the character of $S_\theta$ is $N_S(-1 + 2\cos\theta)$. The polar vector corresponding to the overall translational motion transforms in the same manner and has a contribution of $(-1 + 2\cos\theta)$. The vector corresponding to small angular displacement no more transforms as a polar vector since there is reflection involved in $S_\theta$. Under reflection, the angular displacement transforms are an axial vector. For such a vector, it is well known that the components of the vector parallel to the plane of reflection reverse direction while the component of the vector perpendicular to the plane remains unchanged. After carrying out a rotation by $\theta$, the components of the angular displacement transform as above. Since the plane of reflection is the $x$-$y$ plane after reflection in this plane the final transformations are given by

$$x_a \mapsto -x_a \cos\theta - y_a \sin\theta,$$
$$y_a \mapsto x_a \sin\theta - y_a \cos\theta,$$
$$z_a \mapsto z_a.$$

The contribution to the character from the transformation of a small angular displacement of the molecule by $S_\theta$ is therefore $(1 - 2\cos\theta)$. Adding this contribution to the contribution from the translational motion, and subtracting from the overall, the character of $S_\theta$ in the vibrational representation $\Gamma^V$ is obtained to be

$$\chi^{\Gamma^V}(S_\theta) = N_S(-1 + 2\cos\theta). \tag{4.4.19}$$

It is also clear that $N_S$ can be either 0 or 1. A reflection in the $x$-$y$ plane $\sigma$ is simply a rotary reflection transformation with $\theta = 0$. Its character is therefore

$$\chi^{\Gamma^V} = N_\sigma, \tag{4.4.20}$$

where $N_\sigma$ is the number of the nuclei in the reflection plane. The same formula holds for $\sigma_v$ reflections. For inversions, take $\theta = \pi$ in Equation 4.4.19.

**Example 38.** Consider the methane molecule ($CH_4$). In the equilibrium state of the molecule, the hydrogen atoms are at the vertices of a regular tetrahedron and the carbon is at the geometrical center. Clearly this molecule has $T_d$ point group symmetry. The reader may refer to the character table for the $T_d$ group given in Section 4.2. The molecule has 9 vibrational degrees of freedom. The character of the identity element $E$ is therefore 9. On a $C_3$ axis, there are two nuclei (one hydrogen and the central carbon) which are therefore unmoved by $C_3$ rotations. In accordance with Equation 4.4.18, the character of $C_3$ rotations is zero. The $C_2$ and the $S_4$ axes leave only the central carbon unmoved and the respective characters are 1 and $-1$ (Equation 4.4.19). The characters of $\sigma_d$ reflections are 3 as the $\sigma_d$ planes contain two hydrogen atoms and the central carbon. To summarize

| $T_d$ | $E$ | $8C_3$ | $3C_2$ | $6S_4$ | $6\sigma_d$ |
|---|---|---|---|---|---|
| $\Gamma^V$ | 9 | 0 | 1 | $-1$ | 3 |

Decomposing the vibrational representation, one has

$$\Gamma^V = \mathbf{A_1} \oplus \mathbf{E} \oplus \mathbf{F_2} \oplus \mathbf{F_2}.$$

There are a total of 4 distinct vibrational frequencies,[1] two of which ($\mathbf{F_2}$) are triply degenerate, one is doubly degenerate ($\mathbf{E}$) and one is non-degenerate ($\mathbf{A_1}$). □

# Exercises

1. Suppose we confine a particle to move in two dimensions where the potential is $U(x, y) = 0$ in a region enclosing an equilateral triangle and zero elsewhere. Purely from group theory, deduce whether the stationary states are non-degenerate or degenerate.

2. Suppose the $C_{3v}$ symmetry of a molecule breaks down to a subgroup $C_s = \{E, \sigma_v\}$ due to external perturbation. How does the degree 2 degenerate level of $C_{3v}$ split with respect to the $C_s$ group.

---

[1] https://www.youtube.com/watch?v=3RqEIr8NtMI& index=29& list=PLEB476ECA0DA9481C.

3. Determine the irreducible representations corresponding to the normal modes of the ammonia molecule.

4. Write down the decompositions of all possible tensor products of two irreducible representations of $C_{3v}$. Show that the electric and magnetic dipole transitions between states of representations **A₁** and **A₂** are forbidden.

5. From Equation 4.4.19, calculate the character of the inversion operation $I$.

6. For the symmetry group $D_{3d}$, find all the allowed electric and magnetic dipole transitions. The character table for $D_{3d}$ is produced below for reference.

| $D_{3d}$ | $E$ | $2C_3$ | $3C_2$ | $I$ | $2S_6$ | $3\sigma_d$ | | |
|---|---|---|---|---|---|---|---|---|
| $A_{1g}$ | 1 | 1 | 1 | 1 | 1 | 1 | | $x^2+y^2, z^2$ |
| $A_{2g}$ | 1 | 1 | −1 | 1 | 1 | −1 | $R_z$ | |
| $E_g$ | 2 | −1 | 0 | 2 | −1 | 0 | $(R_x\ R_y)$ | $(x^2-y^2, xy)$ |
| $A_{1u}$ | 1 | 1 | 1 | −1 | −1 | −1 | | |
| $A_{2u}$ | 1 | 1 | −1 | −1 | −1 | 1 | $z$ | |
| $E_u$ | 2 | −1 | 0 | −2 | 1 | 0 | $(x,y)$ | |

7. Find the allowed electric quadrupole transitions for the symmetry group $D_{3d}$.

8. Exploiting the $C_{3v}$ symmetry of an equilateral molecule made of three identical nuclei, determine the degree of degeneracy of the eigenfrequencies. Calculate the eigenfrequencies of small oscillations of the molecule by utilizing the symmetry of the normal modes. Verify that the frequencies are the same as those obtained in Example 35. The bonds connecting the nuclei, all have same spring constant $\varkappa$.

9. Using discrete group symmetry, analyze the normal modes of a square system of four identical masses. Identical springs connect two masses along the edges of the square. The mass points are constrained to move in a plane.

# 5

# Lie Groups and Lie Algebras

The previous chapters dealt with some systems whose symmetry could be described by finite groups. Such systems possess *discrete symmetry*. One does not need to go too far afield to find an example of a system whose symmetry is not discrete. Consider a sphere which appears the same when viewed from all orientations. If the sphere were to be rotated by any angle whatsoever about an axis that passes through the sphere's center, it would still appear the same. Any diametrical plane is also a reflection plane of symmetry. Therefore, there are an infinite number of symmetrical transformations of the sphere. Apart from this fact, it is possible to develop the notion of *closeness* between certain symmetry transformations. If $C_\alpha$ is the symmetry transformation corresponding to a rotation by $\alpha$ about the axis $C$ passing through the sphere's center, then $C_{\alpha+\epsilon}$ is also a symmetry transformation for the sphere where $\epsilon$ could be made arbitrarily small in magnitude. This is an instance of *continuous symmetry*. The theory of Lie groups is the suitable tool for study of such symmetries. It is the purpose of this chapter to introduce the basic concepts of the theory of Lie groups and their associated Lie algebras so as to be able to quickly apply these for problem solving.

## 5.1  The Circle Group U(1)

Consider the group of complex numbers of unit modulus with the group operation being the usual multiplication. All such numbers lie on the unit circle centered at the origin in the complex plane (Figure 5.1.1). This group is called the **U(1)** group. Apart from satisfying the group axioms, **U(1)** has a richer structure owing to the fact that it is a subset of the complex plane. For any two complex numbers $z_1$ and $z_2$, their product $z_1 z_2$ is uniquely defined. Additionally, for small variations in $z_1$ or $z_2$ or both, the value of the product varies by a small amount. More precisely, the map $(z_1, z_2) \to z_1 z_2$ is continuous. The inversion operation $z \to \frac{1}{z}$ for $z \neq 0$ can be shown to be a continuous map. As $\mathbf{U(1)}$ is a subset of the complex plane, these properties are inherited in $\mathbf{U(1)}$. The product and inversion operations in $\mathbf{U(1)}$ are continuous operations and this makes $\mathbf{U(1)}$ a *continuous group*.

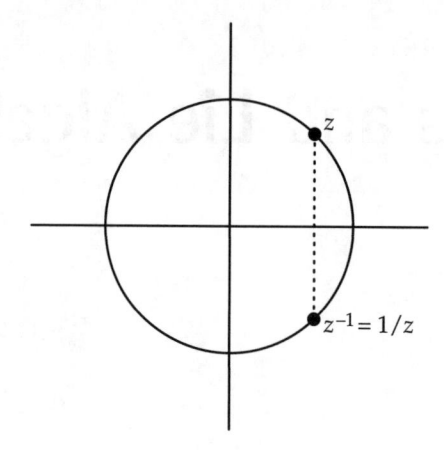

**Figure 5.1.1**  The Circle Group U(1)

Another important feature of $\mathbf{U(1)}$ is that there is a homomorphism from the additive group of real numbers onto $\mathbf{U(1)}$. This homomorphism is clearly given by the map

$$t \to \exp(i\omega t)$$

where $\omega$ is any non-zero real constant, and $t$ varies over the real line $\mathbb{R}$. Thus $\mathbf{U(1)}$ is a *one-parameter group* in the sense that all the group elements are parametrized using a single real parameter $t$. It may be noticed that such a parametrization is not unique and depends on the choice of the value of $\omega$. The identity of $\mathbf{U(1)}$ is prametrized by all integral multiples of $2\pi/\omega$. For group elements $z_1$ and $z_2$ parametrized by $t_1$ and $t_2$, their product is parametrized by $t_1 + t_2$. If $z(t) = z_1 z_2$ and $t$ is assumed to be functionally related to $t_1$ and $t_2$ such as $t = f(t_1, t_2)$, then it is clear that in case of $\mathbf{U(1)}$ this functional dependence is given by

$$f(t_1, t_2) = t_1 + t_2.$$

Likewise, if the parameters of $z$ and $z^{-1}$ are assumed to be related by the function $g(t)$, then in case of $\mathbf{U(1)}$

$$g(t) = -t.$$

Both $f$ and $g$ are not merely continuous but are in fact *analytic functions* (i.e., they can be expanded as a Taylor Series about any point in their domains of definition). $U(1)$ is an almost trivial example of a class of groups called *Lie groups*.

## 5.2 The Matrix Exponential

It is clear from the above discussion that any given parametrization $\exp(i\omega t)$ of $U(1)$ is a one-dimensional unitary representation of the group $U(1)$. The generalisation of such a parametrization and exponentiation for other Lie group elements involving higher dimensional matrices will be

$$g(\{t^a\}) = \exp(i \sum_{a=1}^{s} X_a t^a) \equiv \exp(\Omega) \text{ where } g(\{t^a = 0\}) = \mathbb{I}, \quad (5.2.1)$$

and $t^a$'s are the parameters. Note that the number of parameters depends on the group properties of the Lie group. For instance, a set of $2 \times 2$ unitary matrices forms a group $U(2)$. Though there are 8 real entries (4 complex entries) in these matrices, the unitarity property imposes four real constraints. Hence, the number of independent real parameters to describe these matrices reduces from 8 to 4. Associated with every real parameter, there is a corresponding matrix $X_a$. Hence there will be four linearly independent $X_a$'s for $U(2)$ group. The role of $X_a$'s is to take group elements away from identity $\mathbb{I}$ and hence they are called **generators** of the Lie group.

### 5.2.1 Generators of the Lie group

Suppose $u$ is a real parameter and $\Omega \equiv i \sum_a X_a t^a$ a square matrix. Let $f(u) = \exp(\Omega u)$ be a matrix valued function of $u$. Following the usual rules of differentiation of functions of a real variable, one has

$$\frac{d}{du} f(u) = \exp(\Omega u) \Omega = \Omega \exp(\Omega u). \quad (5.2.2)$$

Such a differentiation cannot determine each of the generators separately. In order to determine the generators, we need to take the parameters $\delta t^a$'s to be infinitesimal so that differentiation with respect to $\delta t^a$ is meaningful; leading to:

$$X_a = -i \frac{\partial}{\partial \delta t^a} \exp\left(i \sum_b \delta t^b X_b\right) \Big|_{\delta t^a = 0}. \quad (5.2.3)$$

Using the group multiplication law, $g(\{t^a\}) g(\{u^b\}) = g(\{v^c(t^a, u^b)\})$, it is a straightforward exercise (as discussed in Section 5.5) to show that the generators $X^a$'s satisfy

$$X_a X_b - X_b X_a = i C_{ab}^c X_c \tag{5.2.4}$$

where coefficients $C_{ab}^c$ are known as structure constants and the above relation is called Lie algebra $\mathfrak{g}$ whose formal definitions and properties are elaborated in Section 5.3.

### 5.2.2 Convergence property of matrix exponentials

It is not at all clear whether the exponential of the square matrix which can also be formally written as the power series expansion

$$\exp \Omega = \sum_{k=0}^{\infty} \frac{\Omega^k}{k!}, \tag{5.2.5}$$

is a convergent series. In the case of $\mathbf{U(1)}$, $\Omega = [i\omega t]$ is a $1 \times 1$ matrix, and the above expansion is same as that of $\exp(i\omega t)$. However, for a general square matrix $\Omega$, Equation 5.2.5 is meaningful only if the infinite series in its right hand side converges. Noting that all the summands are matrices of same order, the critierion for convergence can be set simply to be that all the series of corresponding matrix entries in the summands converge. We will now illustrate this convergence property through a simple example.

**Example 39.** Consider the matrix

$$\Omega \equiv i\omega\theta = \begin{pmatrix} 0 & -\theta \\ \theta & 0 \end{pmatrix}.$$

It is easily seen that

$$\Omega^{2s} = \begin{pmatrix} (-1)^s \theta^{2s} & 0 \\ 0 & (-1)^s \theta^{2s} \end{pmatrix}$$

$$\Omega^{2s+1} = \begin{pmatrix} 0 & (-1)^{s+1} \theta^{2s+1} \\ (-1)^s \theta^{2s+1} & 0 \end{pmatrix} \qquad s = \{0, 1, 2, \ldots\}.$$

With the definition $\exp \Omega \equiv \exp i\omega\theta$ as in Equation 5.2.5, one has

$$\exp \Omega \equiv \exp i\omega\theta = \begin{pmatrix} \cos\theta & -\sin\theta \\ \sin\theta & \cos\theta \end{pmatrix}.$$

The reader should immediately recognize $\exp \Omega$ as the familiar rotation matrix $\mathbf{R}_\theta$ about an axis perpendicular to the $x$-$y$ plane. □

The inverse of such a matrix will be given by $\mathbf{R}_\theta^{-1} \equiv \mathbf{R}_{-\theta} = R_\theta^T$ implying that the matrix $\mathbf{R}_\theta$ denotes an orthogonal matrix with the corresponding $\Omega$ matrix being the antisymmetric matrix $\Omega^T = -\Omega$. Note that the equivalent expression in terms of the $\omega$ matrix will imply that the matrices $\exp(i\omega\theta)$ are unitary with $\omega$ being self-adjoint as illustrated below.

The adjoint of a linear transformation was defined in Section 3.1. If $\Omega \equiv i\omega\theta$ is a square antisymmetric matrix, then it follows from the definition of adjoint that $[(i\omega)^\dagger]_{jk} = \overline{[(i\omega)_{kj}]} = -i\overline{[\omega_{kj}]}$. Further, if $\omega$ itself was self-adjoint then

$$[(i\omega)^\dagger]_{jk} = -i\omega_{jk}$$

$$\Rightarrow (i\omega)^\dagger = -i\omega.$$

Assuming that the matrix exponential $\exp(i\omega\theta)$ converges for some self-adjoint $\omega$, upon taking the adjoint of Equation 5.2.5 one has

$$[\exp(i\omega\theta)]^\dagger = \sum_{k=0}^{\infty} \frac{\theta^k}{k!}[(i\omega)^k]^\dagger = \sum_{k=0}^{\infty} \frac{\theta^k}{k!}(-i\omega)^k$$

$$\Rightarrow [\exp(i\omega\theta)]^\dagger = \exp(-i\omega\theta).$$

From the definition of the matrix exponential, it can be shown that

$$\exp(-i\omega\theta)\exp(i\omega\theta) = \exp(i\omega\theta)\exp(-i\omega\theta) = \mathbb{I}. \tag{5.2.6}$$

In other words, $\exp(i\omega\theta)$ is a unitary matrix if $\omega$ is self-adjoint. This property has several applications in quantum mechanics as will be seen in the next chapter. It is important to bear in mind that for two square matrices $\Omega_1$ and $\Omega_2$ of same order, the equality $\exp\Omega_1 \exp\Omega_2 = \exp(\Omega_1 + \Omega_2)$ holds only if the $\Omega_1$ and $\Omega_2$ commute. Such a property is obeyed by the set of rotation matrices $\{\mathbf{R}_\theta\}$ which are orthogonal $2 \times 2$ matrices. Incidentally, these $2 \times 2$ matrices, involving four real entries, are dependent only on one parameter $\theta$. This is because the orthogonality property gives three constraints reducing four real entries of $2 \times 2$ matrices to one. Hence this set $R_\theta$ forms a group $SO(2)$ where the letter $O$ denotes the matrices are orthogonal and $S$ implies that their determinants are unity. Like $U(1)$, the one parameter group $SO(2)$ is also abelian group:

$$R_{\theta_1} R_{\theta_2} = R_{\theta_2} R_{\theta_1} \tag{5.2.7}$$

The *commutator* of $\Omega_1$ and $\Omega_2$, denoted by the symbol $[\Omega_1, \Omega_2]$, is defined as

$$[\Omega_1, \Omega_2] = \Omega_1\Omega_2 - \Omega_2\Omega_1. \tag{5.2.8}$$

Thus the condition for $\exp\Omega_1 \exp\Omega_2 = \exp(\Omega_1 + \Omega_2)$ to be true is that $[\Omega_1, \Omega_2] = 0$. The following properties of the commutator are directly obtained from its definition.

(1) $[\lambda\Omega_1 + \mu\Omega_2, \Omega_3] = \lambda[\Omega_1, \Omega_3] + \mu[\Omega_2, \Omega_3]$ for any complex $\lambda$ and $\mu$.

(2) $[\Omega_1, \Omega_2] = -[\Omega_2, \Omega_1]$.

(3) $[\Omega_1, [\Omega_2, \Omega_3]] + [\Omega_2, [\Omega_3, \Omega_1]] + [\Omega_3, [\Omega_1, \Omega_2]] = 0$.

The last equality is called the *Jacobi identity*. All the three aforementioned properties of the commutator are also properties of the *Poisson bracket* if $\Omega_i$ were dynamical quantities of a Hamiltonian describing classical system. The generators of a Lie group (5.2.4) indeed have such non-trivial commutators. We will formally introduce the algebra obeyed by the generators $X_a$'s in the following section.

## 5.3 Finite Dimensional Lie Algebras

**Definition 5.** A *Lie Algebra* $\mathfrak{g}$ is a vector space on which is defined a binary operation [ , ] having following properties:

(1) For all $x$ and $y$ in $\mathfrak{g}$, $[x, y]$ is in $\mathfrak{g}$.
(2) For all $x$, $y$ and $z$ in $\mathfrak{g}$, and scalars $\lambda$ and $\mu$, $[\lambda x + \mu y, z] = \lambda[x, z] + \mu[y, z]$.
(3) $[x, y] = -[y, x]$.
(4) $[x, [y, z]] + [y, [z, x]] + [z, [x, y]] = 0$. [Jacobi identity]

If the scalars are chosen from the set of real numbers, then the Lie algebra is called a *real Lie algebra*. In case the scalars come from the set of complex numbers, one has a *complex Lie algebra*. The properties of the composition [ , ] are identical to the properties of the commutator discussed in the last section. The composition $[x, y]$ of $x, y \in \mathfrak{g}$ would be refered to as the *Lie bracket* of $x$ and $y$. A Lie algebra $\mathfrak{g}$ in which the Lie bracket vanishes identically over $\mathfrak{g}$ is an *abelian algebra*. Being a vector space, $\mathfrak{g}$ has a basis. It will be assumed throughout that $\mathfrak{g}$ is finite dimensional. If $\{x_i\}_{i=1}^n$ is a basis for $\mathfrak{g}$ then for any $x_s$ and $x_t$ in the basis it must be true that

$$[x_s, x_t] = \sum_{k=1}^n c_{st}^k x_k \qquad (5.3.1)$$

since $[x_s, x_t]$ is after all yet another vector in $\mathfrak{g}$. The coefficients $c_{st}^k$ are called the *structure constants* of the Lie algebra. The obvious relation $[x_s, x_t] + [x_t, x_s] = 0$ for commutator bracket puts the following restriction on the structure constants:

$$\sum_{k=1}^n (c_{st}^k + c_{ts}^k) x_k = 0.$$

From linear independence of $\{x_i\}_{i=1}^n$, it follows that $c_{st}^k = -c_{ts}^k$. In a Lie algebra, it can be shown further that

$$\sum_{k=1}^n [c_{sk}^l c_{tu}^k + c_{tk}^l c_{us}^k + c_{uk}^l c_{st}^k] = 0$$

# Lie Groups and Lie Algebras

The following definitions of Lie algebra and the properties of the structure functions are essential and will henceforth be assumed for the rest of the chapter. We will now present two familiar examples where the structure constants and Lie algebra appear.

**Example 40.** The cross product of vectors in the three-dimensional space $\mathbb{R}^3$ turns into a three-dimensional Lie algebra. If the basis vectors are taken to be the standard **i**, **j** and **k**, then all the four properties of Definition 5 are easily seen to be satisfied. The non-zero structure constants are

$$c^i_{jk} = c^j_{ki} = c^k_{ij} = +1$$

$$c^i_{kj} = c^j_{ik} = c^k_{ji} = -1,$$

and the other structure constants are all equal to zero. □

**Example 41.** In quantum mechanics, the operators of position $\hat{x}$ and the corresponding component of momentum $\hat{p}_x$ of a particle do not commute and the identities

$$[\hat{x}, \hat{p}_x] = [\hat{y}, \hat{p}_y] = [\hat{z}, \hat{p}_z] = i\hbar \tag{5.3.2}$$

are well known. For simplicity, let us set $\hbar$ (proportional to Planck constant) to be 1. The position components and the momentum components commute amongst themselves. Also, position components commute with any orthogonal momentum component. The components of the orbital angular momentum operators $\hat{\ell} = \hat{\mathbf{r}} \times \hat{\mathbf{p}}$ of the particle

$$\hat{\ell}_x = \hat{y}\hat{p}_z - \hat{z}\hat{p}_y$$

$$\hat{\ell}_y = \hat{z}\hat{p}_x - \hat{x}\hat{p}_z$$

$$\hat{\ell}_z = \hat{x}\hat{p}_y - \hat{y}\hat{p}_x,$$

generate a three-dimensional Lie algebra. Using the position-momentum algebra (5.3.2) and the definitions of the commutator, the readers can verify that the Lie algebra amongst the angular momentum components become:

$$[\hat{\ell}_x, \hat{\ell}_y] = i\hat{\ell}_z$$

$$[\hat{\ell}_y, \hat{\ell}_z] = i\hat{\ell}_x \tag{5.3.3}$$

$$[\hat{\ell}_z, \hat{\ell}_x] = i\hat{\ell}_y \qquad \square$$

We will now focus on rotation operations in three-dimensional space and apply the above algebra. We can consider three independent rotations depending on the choice of perpendicular axis. Let $R(\boldsymbol{\theta} \equiv \theta\hat{n})$ denote $3 \times 3$ matrix representation of proper rotations $\theta$ about $\hat{n}$ axis, acting on any position vector $\vec{r}_1$ as follows:

$$R(\boldsymbol{\theta}) : \vec{r}_1 \to \vec{r}_2,$$

where rotations keeps the norm of vectors $\vec{r}_1.\vec{r}_1 = \vec{r}_2.\vec{r}_2$ same. This implies that the rotation $R(\theta)$ must obey the orthogonal property

$$R(\boldsymbol{\theta})R^T(\boldsymbol{\theta}) = \mathbb{I}.$$

The determinant of these proper rotations matrices must be one. Depending on the axis of rotation, there will be an appropriate $\Omega \equiv -i\boldsymbol{L}.\boldsymbol{\theta}$ in the exponential map: $R(\theta) = \exp(-i\boldsymbol{\theta}.\boldsymbol{L})$. Just like the collection of one parameter elements $\{\exp(i\omega t)\}$ of $U(1)$ group and $\{\exp(i\omega\theta)\}$ of $SO(2)$ group, the set of orthogonal $3 \times 3$ matrices $R(\boldsymbol{\theta}) = \exp(\Omega) = \exp(-i\boldsymbol{\theta}.\boldsymbol{L})$, involving three parameters $\boldsymbol{\theta}$ and three generators $\boldsymbol{L}$, forms a group $SO(3)$. We leave it to the readers to verify that orthogonal $3 \times 3$ matrices will require only three independent entries. Unlike the $SO(2)$ group (5.2.7) where the elements commute, the $SO(3)$ group elements do not obey

$$\exp(i\Omega_1)\exp i\Omega_2 \neq \exp[i(\Omega_1 + \Omega_2)]$$

and hence the group is non-abelian. For infinitesimal angle $\delta\theta$, the exponential map of $SO(3)$ elements can be approximated as

$$R(\delta\boldsymbol{\theta}) = \mathbb{I} - i\delta\theta\hat{n}.\boldsymbol{L}, \qquad (5.3.4)$$

where we have denoted the magnitude of the rotation angle about an axis $\hat{n}$

**Example 42.** It is straightforward to check that the infinitesimal rotations of magnitude $\delta\theta$ about x-axis $R(\delta\theta\hat{i})$ and the infinitesimal rotations of same magnitude $\delta\theta$ about y-axis $R(\delta\theta\hat{j})$ obey the following relation:

$$R(\delta\theta\hat{i})R(\delta\theta\hat{j}) - R(\delta\theta\hat{j})R(\delta\theta\hat{i}) = R(\delta\theta^2\hat{k}) - \mathbb{I}. \qquad (5.3.5)$$

Using the definition (5.3.4) in the above equation, verify that the $3 \times 3$ matrices $L_x, L_y, L_z$ obey angular momentum algebra (5.3.3):

$$[L_x, L_y] = iL_z.$$

Using $R(\delta\theta\hat{i}) \equiv I - i\delta\theta L_x$ and $R(\delta\theta\hat{j}) \equiv I - i\delta\theta L_y$, the LHS (5.3.5) simplifies to $(-i)^2\delta\theta^2[L_x, L_y]$ whereas the RHS (5.3.5) gives $-i\delta\theta^2 L_z$ leading to the conventional angular momentum algebra (5.3.3). $\square$

Thus $L_x, L_y, L_z$, which are the generators of the rotation group $SO(3)$, constitute the Lie algebra $\mathfrak{g}$ of the rotation group $SO(3)$ denoted as $\mathfrak{g} \equiv \mathfrak{so}(3)$ which is isomorphic

# Lie Groups and Lie Algebras

to the angular momentum algebra (5.3.3). Incidentally, the range of the magnitude of the parameter $\theta = \theta\hat{n}$ must be $-\pi \leq \theta < \pi$. Hence the parameter space describing the group $SO(3)$ will be a solid sphere of radius $\pi$. The groups with such bounded parameters are referred to as *compact groups*.

Every point inside the solid sphere corresponds to a distinct group element of $SO(3)$. However, the diametrically opposite points on the boundary of the solid sphere of radius $\pi$ will denote the same group element $(R[\pi\hat{n}] = R[\pi(-\hat{n})])$. Hence, this identification of diametrically opposite points allows two topologically distinct closed curves inside the sphere as shown in Figure 5.3.1. In the literature, such a parameter space is called *doubly-connected space*.

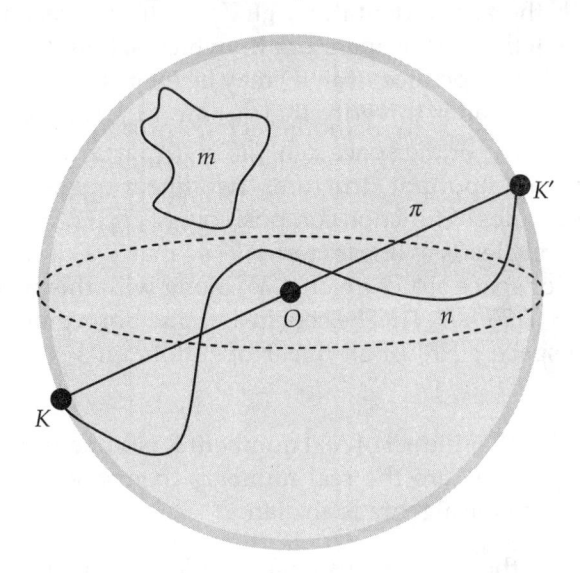

**Figure 5.3.1** Group Manifold of the $SO(3)$ (doubly-connected space)

**Definition 6.** If $\mathfrak{g}$ is a Lie algebra and $\mathfrak{h}$ is a subset of it such that the elements of $\mathfrak{h}$ form a Lie algebra under the Lie bracket of $\mathfrak{g}$, then $\mathfrak{h}$ is called a *subalgebra*. The subalgebra $\mathfrak{h}$ is an *invariant subalgebra* if $[g, h] \in \mathfrak{h}$ for all $g \in \mathfrak{g}$ and $h \in \mathfrak{h}$.

**Example 43.** Suppose $g \in \mathfrak{g}$ and $h \in \mathfrak{h}$ where $\mathfrak{h}$ is an invariant subalgebra of $\mathfrak{g}$. Show that the set of elements $\{\exp(ih)\}$ forms an invariant subgroup $H \subset G$ where the set $\{\exp(ig)\}$ belongs to Lie group G. In other words, show that

$$\exp(ig)\exp(ih)\exp(-ig) = \exp(ih_1)$$

where $h_1 \in \mathfrak{h}$.

Using the well-known properties of matrices and the fact that $h \in \mathfrak{h}$, we can deduce

$$\exp(ig)h\exp(-ig) = h + i[g, h] - \frac{1}{2!}[g, [g, h]] + \ldots = h_1,$$

as $[g, h]$, $[g, [g, h]]$, ... $\in \mathfrak{h}$. Hence, for any exponential form of $h$ in LHS of the above equation, we will get RHS as $\exp(ih_1)$. □

In Chapter 3, linear representations of finite groups were studied in some length. Analogously, representations of a Lie algebra may be defined. To begin with, let $V$ be a complex (or real) vector space of dimension $n$. Let $\mathfrak{gl}(V)$ be the set of all linear operators on $V$. For $T_1, T_2 \in \mathfrak{gl}(V)$ and scalars $\alpha$, $\beta$ one may define a linear combination $\alpha T_1 + \beta T_2$ to be the linear operator which acts on $V$ in accordance with

$$(\alpha T_1 + \beta T_2)x = \alpha T_1(x) + \beta T_2(x),$$

for all $x$ in $V$. With the above definition, $\mathfrak{gl}(V)$ itself is a vector space. If $\{x_i\}_{i=1}^n$ is a basis for $V$, then define linear operators $X_{ij}$ which act on the basis of $V$ such that $X_{ij}(x_k) = \delta_{ik} x_j$. Any linear operator in $\mathfrak{gl}(V)$ may be expressed as a linear combination of $X_{ij}$'s. It is left to the reader to verify that $X_{ij}$ are in fact linearly independent. It follows that the dimension of the space $\mathfrak{gl}(V)$ is $n^2$. Apart from being a vector space, $\mathfrak{gl}(V)$ has an additional important structure. Two linear operators can be composed according to the usual rules of function composition, $(T_1 T_2)(x) = T_1(T_2(x))$. Linearity of $T_1$ and $T_2$ immediately leads to linearity of $T_1 T_2$ so that $T_1 T_2$ is also in $\mathfrak{gl}(V)$. One can now employ the vector space structure of $\mathfrak{gl}(V)$ along with this property to define the Lie bracket $[T_1, T_2] = T_1 T_2 - T_2 T_1$. The conclusion that, for a given finite dimensional vector space $V$, the space $\mathfrak{gl}(V)$ of all linear operators on $V$ can be turned into Lie algebra is obvious.

**Example 44.** It is trivial that the set of real numbers $\mathbb{R}$ is a one-dimensional real vector space. The linear operators are the real numbers themselves and the commutator identically vanishes. The Lie algebra is abelian. □

**Example 45.** Consider the two-dimensional complex vector space. In any basis, the linear operators on this space can be represented by $2 \times 2$ matrices with complex entries. The Lie algebra of this four-dimensional complex vector space of linear operators is denoted $\mathfrak{gl}(2, \mathbb{C})$. $\mathfrak{gl}(2, \mathbb{C})$ is clearly non-abelian which is spanned by four $2 \times 2$ matrices such as

$$E_1 = \begin{bmatrix} 1 & 0 \\ 0 & 0 \end{bmatrix}, \quad E_2 = \begin{bmatrix} 0 & 1 \\ 0 & 0 \end{bmatrix},$$

$$E_3 = \begin{bmatrix} 0 & 0 \\ 1 & 0 \end{bmatrix}, \quad E_4 = \begin{bmatrix} 0 & 0 \\ 0 & 1 \end{bmatrix}.$$

**Example 46.** There are several important subalgebras of $\mathfrak{gl}(2, \mathbb{C})$. For instance, consider matrices of the form

$$X = \begin{bmatrix} a & z \\ z^* & -a \end{bmatrix},$$

where $a$ is a real number and $z$ a complex number. The important properties of such matrices are that they are traceless and Hermitian. The complex subspace of $\mathfrak{gl}(2, \mathbb{C})$ spanned by matrices of this type is a subalgebra denoted by $\mathfrak{sl}(2, \mathbb{C})$. If $\{X_i\}_{i=1}^n$ are traceless Hermitian matrices and $\{\alpha_i\}_{i=1}^n$ complex numbers, then $X = \sum_{i=1}^n \alpha_i X_i$ is in $\mathfrak{sl}(2, \mathbb{C})$. Similarly, for $Y = \sum_{j=1}^m \beta_j Y_j$ in $\mathfrak{sl}(2, \mathbb{C})$, by linearity of the Lie bracket, one has

$$[X, Y] = \sum_{i,j} \alpha_i \beta_j [X_i, Y_j] \equiv Z.$$

The reader can verify that the brackets $[X_i, Y_j]$ are all equal to the imaginary unit times $Z_{ij}$, where $Z_{ij}$ is again a traceless Hermitian matrix. Then the above equality shows that $[X, Y] \in \mathfrak{sl}(2, \mathbb{C})$. Hence, $\mathfrak{sl}(2, \mathbb{C})$ is indeed a subalgebra of $\mathfrak{gl}(2, \mathbb{C})$. Since $\mathfrak{sl}(2, \mathbb{C})$ is a proper subspace of $\mathfrak{gl}(2, \mathbb{C})$, its dimension is less than 4. Notice that a traceless Hermitian matrix is completely specified if the 3 independent quantities $a$, $z$ and $z^*$ are known. Therefore the dimension of $\mathfrak{sl}(2, \mathbb{C})$ must be 3. This can be seen more explicitly by considering the matrices

$$\sigma_x = \begin{bmatrix} 0 & 1 \\ 1 & 0 \end{bmatrix}, \sigma_y = \begin{bmatrix} 0 & -i \\ i & 0 \end{bmatrix}, \sigma_z = \begin{bmatrix} 1 & 0 \\ 0 & -1 \end{bmatrix}.$$

$\sigma_x$, $\sigma_y$ and $\sigma_z$ are traceless Hermitian matrices which are also linearly independent. Their linear span is definitely a subset of $\mathfrak{sl}(2, \mathbb{C})$. It follows that their linear span is all of $\mathfrak{sl}(2, \mathbb{C})$. The matrices $\sigma_x$, $\sigma_y$ and $\sigma_z$ are the familiar Pauli matrices.

Complex numbers $z$ in $\mathbb{C}$ are pairs of independent real numbers $(\Re(z), \Im(z))$. Similarly, an $n$-tuple of complex numbers is a $2n$-tuple of real numbers. Thus a complex vector space of dimension $n$ can be regarded as a real vector space of dimension $2n$. In the above discussion, $\mathfrak{gl}(2, \mathbb{C})$ was regarded as a complex vector space of dimension 4. As a real vector space, $\mathfrak{gl}(2, \mathbb{C})$ has dimension 8 spanned by the basis consisting of $E_1$, $E_2$, $E_3$ and $E_4$ and the matrices

$$\begin{bmatrix} i & 0 \\ 0 & 0 \end{bmatrix}, \begin{bmatrix} 0 & i \\ 0 & 0 \end{bmatrix}$$

$$\begin{bmatrix} 0 & 0 \\ i & 0 \end{bmatrix}, \begin{bmatrix} 0 & 0 \\ 0 & i \end{bmatrix}.$$

Likewise, $\mathfrak{sl}(2, \mathbb{C})$ is a six-dimensional real Lie algebra spanned by a basis consisting of $\sigma_x$, $\sigma_y$, $\sigma_z$ and the matrices

$$S_x = \begin{bmatrix} 0 & i \\ i & 0 \end{bmatrix}, S_y = \begin{bmatrix} 0 & 1 \\ -1 & 0 \end{bmatrix}, S_z = \begin{bmatrix} i & 0 \\ 0 & -i \end{bmatrix}.$$

The following Lie brackets are easily computed from the above *representations* of the basis elements of $\mathfrak{sl}(2, \mathbb{C})$.

$$[\sigma_x, \sigma_y] = 2i\sigma_z \quad [\sigma_y, \sigma_z] = 2i\sigma_x \quad [\sigma_z, \sigma_x] = 2i\sigma_y \tag{5.3.6}$$
$$[S_x, S_y] = -2S_z \quad [S_y, S_z] = -2S_x \quad [S_z, S_x] = -2S_y.$$

The lower set of 3 equations is completely equivalent to the upper 3 equations since $S_x = i\sigma_x$, etc. The three-dimensional real subalgebra of $\mathfrak{sl}(2, \mathbb{C})$ spanned by $S_x$, $S_y$ and $S_z$ is called the $\mathfrak{su}(2)$ algebra. Notice that $S_x$, $S_y$ and $S_z$ are skew-Hermitian ($S_x^\dagger = -S_x$ etc.) traceless matrices. □

**Definition 7.** Let $\mathfrak{g}$ be a Lie algebra and $\mathfrak{gl}(V)$ the Lie algebra of operators on the vector space $V$. A linear map $\pi$ from $\mathfrak{g}$ into $\mathfrak{gl}(V)$ is said to be a representation of $\mathfrak{g}$ on the vector space $V$ if

$$\pi([x, y]) = [\pi(x), \pi(y)] \tag{5.3.7}$$

for all $x, y$ in $\mathfrak{g}$. The dimension of $V$ is the degree of the representation $\pi$.

Since a Lie algebra $\mathfrak{g}$ is also a vector space, the above definition of representation allows one to represent $\mathfrak{g}$ over itself. For some $x \in \mathfrak{g}$, consider a mapping $\mathrm{ad}_x$ from $\mathfrak{g}$ into $\mathfrak{g}$ defined by $\mathrm{ad}_x(y) = [x, y]$ for all $y \in \mathfrak{g}$. It is clear that $\mathrm{ad}_x$ is a linear operator on the vector space $\mathfrak{g}$ since the Lie bracket is linear in $y$ by definition. Therefore $\mathrm{ad}_x \in \mathfrak{gl}(\mathfrak{g})$ and one has a mapping $\mathrm{ad}$ from $\mathfrak{g}$ into $\mathfrak{gl}(\mathfrak{g})$ defined as $\mathrm{ad}(x) = \mathrm{ad}_x$ for all $x$ in $\mathfrak{g}$. This mapping is a representation if $\mathrm{ad}([x, y]) = [\mathrm{ad}(x), \mathrm{ad}(y)]$. Let $z \in \mathfrak{g}$ so that one has

$$\mathrm{ad}_{[x,y]}(z) = [[x, y], z] = [x, [y, z]] + [y, [z, x]]$$

$$\Rightarrow \mathrm{ad}_{[x,y]}(z) = [x, [y, z]] - [y, [x, z]]$$

$$\Rightarrow \mathrm{ad}_{[x,y]} = \mathrm{ad}_x \, \mathrm{ad}_y - \mathrm{ad}_y \, \mathrm{ad}_x \tag{5.3.8}$$

and one has shown that $\mathrm{ad}$ is indeed a representation. Such a representation of a Lie algebra is called an *adjoint representation*. From Equation 5.3.1 it immediately follows that the representing matrices of the bases $\{x_s\}_{i=1}^n$ of $\mathfrak{g}$ are given by

$$[\mathrm{ad}_{x_s}]_{kt} = c_{st}^k. \tag{5.3.9}$$

**Example 47.** The $\mathfrak{su}(2)$ algebra was described in the previous example (Equation 5.3.6). If one takes $S_x/2$, $S_y/2$ and $S_z/2$ as the basis for the algebra then in the adjoint representation, the matrices are easily found to be

$$S_x/2 = \begin{bmatrix} 0 & 0 & 0 \\ 0 & 0 & 1 \\ 0 & -1 & 0 \end{bmatrix}, S_y/2 = \begin{bmatrix} 0 & 0 & -1 \\ 0 & 0 & 0 \\ 1 & 0 & 0 \end{bmatrix},$$

$$S_z/2 = \begin{bmatrix} 0 & 1 & 0 \\ -1 & 0 & 0 \\ 0 & 0 & 0 \end{bmatrix}.$$ □

### 5.3.1 su(2) algebra

In quantum mechanics, we define generators of $su(2)$ Lie algebra as $J_i$'s obeying

$$[J_i, J_j] = i\epsilon_{ijk}J_k.$$

The representation for $J_i = \frac{1}{2}\sigma_i$ is known as the fundamental matrix representation which act on a two-dimensional complex vector space $V$. Using the redefinition $(J_1 \pm iJ_2)/\sqrt{2} = J_\pm$ where $J_\pm$ are known as raising and lowering operators, the algebra becomes

$$[J_+, J_-] = J_3 \,; [J_3, J_\pm] = \pm J_\pm. \tag{5.3.10}$$

In fact such a redefinition naturally appears in the formal discussion of Cartan classification of semi-simple Lie algebras. We will briefly present such ladder operators in Section 5.7.

As the $su(2)$ algebra has no abelian subalgebra, it is always possible to choose one of the generators to have a diagonal matrix representation. Following the convention in many books, we take the representation of $J_3$ as diagonal matrix and work with its eigenvectors as basis $\{|j,m\rangle\} \in V$. That is,

$$J_3|j,m\rangle = m|j,m\rangle,$$

where $m$ can be either integers or half-odd integers. We refer to maximum eigenvalue of $J_3$ as $j$ and the corresponding highest value state is $|j,j\rangle$.

Applying $J_\pm$ on the states $|j,m\rangle$ we observe

$$J_3\{J_\pm|jm\rangle\} = J_\pm\{J_3|j,m\rangle\} \pm J_\pm|j,m\rangle = (m \pm 1)\{J_\pm|j,m\rangle\}.$$

From this relation, it is clear that the $J_3$ eigenvalue of states $J_\pm|jm\rangle$ is $m \pm 1$. For highest value state $|j,j\rangle$, we cannot get $J_3$ eigenvalue as $j+1$ for $J_+|j,j\rangle$ which implies

$$J_+|j,j\rangle = 0.$$

The other eigenbasis of $J_3$ can be obtained by applying lowering operator $J_-$ on the highest value state $|j,j\rangle$:

$$J_-|j,j\rangle = N_{j,j}|j,j-1\rangle.$$

Taking inner product of the above state with its dual state $\langle j,j|J_+$, we get

$$\langle j,j|J_+J_-|j,j\rangle = |N_{j,j}|^2 = \langle j,j|J_3 + J_-J_+|j,j\rangle = \langle j,j|J_3|j,j\rangle = j,$$

implying $N_{j,j} = \sqrt{j}$. We can now determine

$$J_+|j,j-1\rangle = \frac{1}{N_{j,j}}J_+J_-|j,j\rangle = \frac{1}{N_{j,j}}\{J_3 + J_-J_+\}|j,j\rangle = \sqrt{j}|j,j\rangle = N_{j,j}|j,j\rangle.$$

This application of $J_-$ can be continued to obtain the basis states $|j,j-2\rangle$, $|j,j-3\rangle$,... belonging to the vector space $V$:

$$J_-|j,j-r\rangle = N_{j,j-r}|j,j-r-1\rangle \;;\; J_+|j,j-r-1\rangle = N_{j,j-r}|j,j-r\rangle.$$

Using the $\mathfrak{su}(2)$ algebra involving ladder operators, it is easy to deduce

$$N^2_{j,j-r} = \langle j,j-r|J_+J_-|j,j-r\rangle = N^2_{j,j-r+1} + j - r,$$

giving a recursion relation amongst the coefficients whose solution turns out to be

$$N_{j,m} = \frac{1}{\sqrt{2}}\sqrt{(j+m)(j-m+1)}.$$

As the vector space is finite dimensional, we will also have a lowest eigenvalue state $|j,j-k\rangle$ such that

$$J_-|j,j-k\rangle = 0.$$

Clearly, $N_{j,m} = 0$ when $m = j+1$ (highest value state $|j,j\rangle$) and $m = j-k = -j$ (lowest value state $|j,j-k\rangle$) which means $k = 2j$. The highest value $j$ can be also determined by acting $J \cdot J = J_3^2 + J_+J_- + J_-J_+$ on $|j,m\rangle$:

$$J \cdot J|j,m\rangle = (J_3^2 + J_+J_- + J_-J_+)|j,m\rangle = j(j+1)|j,m\rangle.$$

Notice that $J \cdot J$ commutes with the all the three $\mathfrak{su}(2)$ generators.

$$[J \cdot J, J_3] = [J \cdot J, J_+] = [J \cdot J, J_3] = 0.$$

Such quantities are called *Casimir operators*.

We will see that the above discussion on $\mathfrak{su}(2)$ algebra and their action on basis states of finite dimensional vector space $V$ will be useful when we formally introduce root vectors and study $\mathfrak{su}(2)$ subalgebras constituting compact semi-simple Lie algebras in Section 5.7.

By exponential mapping, the special unitary group $SU(2)$ group elements will be

$$g(\theta) = \exp(i\theta.J),$$

whose parameters are similar to that of group $SO(3)$ but there is a subtle difference. For the fundamental representation $J = \sigma/2$, using the properties of the Pauli matrices, these group elements can be written as

$$g(\theta\hat{n}) = \cos(\theta/2)\mathbb{I} + i\hat{n}.\sigma \sin(\theta/2),$$

which implies $g(2\pi) \neq g(0)$, suggesting that the parameter space will not be a solid sphere of radius $\pi$. Instead, the $SU(2)$ parameter space is a solid sphere of radius $2\pi$. Further, all points on the boundary of the solid sphere are identified making this $SU(2)$ parameter space *simply connected*. In other words, there will be only one class of closed curves in the parameter space. This subtle difference in the parameter space also indicates that the group $SO(3)$ (Figure 5.3.1) is not isomorphic to group $SU(2)$. That is, $g(\theta\hat{n})$ and $g((\theta + \pi)\hat{n})$ are mapped to $R(\theta\hat{n}) \in SO(3)$ giving *two to one* mapping (homomorphism). In the literature, $SU(2)$ is refered to as a *double cover* of $SO(3)$ because of the above properties.

## 5.4 Semi-simple Lie Algebras

The explicit construction of the adjoint representation of a Lie algebra allows for introduction of the *Killing form* on it. The advantage of the Killing form lies in that it bears some resemblance to the concept of inner product on an inner product space. One should bear in mind that this resemblance is exact only for a certain type of Lie algebras. It is very convenient that such Lie algebras turn out to be quite useful in physical applications.

Let $\mathfrak{g}$ be a Lie algebra and ad its adjoint representation. For $x, y \in \mathfrak{g}$, the Killing form $\kappa(x, y)$ is given by

$$\kappa(x, y) = -\text{tr}(\text{ad}(x)\,\text{ad}(y)), \tag{5.4.1}$$

where **tr** stands for trace of the matrix products. Evidently $\kappa(x, y) = \kappa(y, x)$. The linearity of the map ad gives $\kappa(\alpha x + \beta y, z) = \alpha \kappa(x, z) + \beta \kappa(y, z)$ for scalars $\alpha$, $\beta$ and $x, y, z \in \mathfrak{g}$. Killing form is therefore a symmetric bilinear form. If either $x$ or $y$ is 0, then $\kappa(x, y) = 0$.

Consider the set $\mathfrak{g}^\perp = \{x \in \mathfrak{g} \,|\, \kappa(x, y) = 0 \,\forall\, y \in \mathfrak{g}\}$. $\mathfrak{g}^\perp$ is clearly a subspace of $\mathfrak{g}$ as follows from linearity of $\kappa$. The notation seems to suggest that $\mathfrak{g}^\perp$ consists of vectors orthogonal to the linear space $\mathfrak{g}$. However, this orthogonality is with respect to the Killing form (5.4.1) and not in the usual sense of orthogonality in an inner product space. It very well might be that $\mathfrak{g}^\perp$ is not merely the null space and $\mathfrak{g}^\perp \subset \mathfrak{g}$. When this is true, the Killing form $\kappa$ is said to be *degenerate*.

**Definition 8.** A Lie algebra whose Killing form is non-degenerate is called *semi-simple*.

Let $\mathfrak{h}$ be an invariant subalgebra (Definition 6) of a semi-simple Lie algebra $\mathfrak{g}$. $\mathfrak{h}^\perp = \{x \in \mathfrak{g} \,|\, \kappa(x, y) = 0 \,\forall\, y \in \mathfrak{h}\}$ is the orthogonal subspace of $\mathfrak{h}$ in $\mathfrak{g}$. Also suppose $g \in \mathfrak{g}$, $h \in \mathfrak{h}$ and $x \in \mathfrak{h}^\perp$ so that $[g, h] \in \mathfrak{h}$ and $\kappa(x, h) = 0$. Then one has $\kappa([x, g], h) = \mathbf{tr}\,(\mathrm{ad}([x, g])\mathrm{ad}(h))$ and from the property of trace and Equation 5.3.8, $\kappa([x, g], h) = \kappa(x, [g, h]) = 0$. It follows that $[x, g] \in \mathfrak{h}^\perp$ and since $g$ is arbitrary in $\mathfrak{g}$, $\mathfrak{h}^\perp$ is an invariant subagebra in $\mathfrak{g}$. The reader can easily verify that $\mathfrak{h} \cap \mathfrak{h}^\perp$ is also an invariant subalgebra in $\mathfrak{g}$. Since $\kappa(x, y) = 0$ for $x, y \in \mathfrak{h}^\perp \cap \mathfrak{h}$, it follows that for any $g \in \mathfrak{g}$, $\kappa(g, [x, y]) = \kappa(x, [y, g]) = 0$ requires $\mathfrak{h}^\perp \cap \mathfrak{h}$ is an abelian subalgebra because $\kappa$ is non-degenerate for $\mathfrak{g}$.

A basis $\{x_i\}_{i=1}^{k}$ for the subspace $\mathfrak{h} \cap \mathfrak{h}^\perp$ can be chosen which can be extended to a complete basis $\{x_i\}_{i=1}^{k+l}$ for $\mathfrak{g}$. $\{x_j\}_{j=k+1}^{k+l}$ is then a basis for the complement of the subspace $\mathfrak{h} \cap \mathfrak{h}^\perp$ in $\mathfrak{g}$. As $\mathfrak{h} \cap \mathfrak{h}^\perp$ is an abelian subalgebra, it immediately follows that the matrix of the operator $\mathrm{ad}\,(g)\,\mathrm{ad}\,(x)$ is of the form

$$\begin{bmatrix} 0_{k\times k} & B_{k\times l} \\ 0_{l\times k} & 0_{l\times l} \end{bmatrix}$$

for any $x \in \mathfrak{h} \cap \mathfrak{h}^\perp$ and $g \in \mathfrak{g}$. The size of the various 0-blocks and the matrix $B$ has been indicated by subscripts. From the above it follows that $\mathbf{tr}\,(\mathrm{ad}(g)\mathrm{ad}(x)) = \kappa(g, x) = 0$ for all $g \in \mathfrak{g}$. Because $\kappa$ is non-degenerate by assumption, the only possible conclusion is $x = 0$ and one has $\mathfrak{h} \cap \mathfrak{h}^\perp = \{0\}$. Hence, for semi-simple Lie algebra, it can be shown that $\mathfrak{g} = \mathfrak{h} \oplus \mathfrak{h}^\perp$.

Suppose $\{h_i\}_{i=1}^{m}$ is a basis of the subspace $\mathfrak{h}$. Any $x \in \mathfrak{h}$ can be expressed $x = \sum_{i=1}^{m} c_i h_i$ where $c_i$ are real or complex (depending upon whether $\mathfrak{g}$ is a real or complex vector space). Thus every $x \in \mathfrak{h}$ corresponds to an $m$-tuple of numbers. Let the collection of all such $m$-tuples be called $F^m$. Here $F$ stands for the *field* $\mathbb{R}$ or $\mathbb{C}$. With component wise addition and scalar multiplication of the $m$-tuples, $F^m$ is a vector space which has the same dimension as the subspace $\mathfrak{h}$. Consider a map $T$ (Section 3.1) from $\mathfrak{g}$ into $F^m$ such that for $g \in \mathfrak{g}$,

$$T(g) = (\kappa(g, h_1), \ldots, \kappa(g, h_m)).$$

$T$ is then a linear transformation which is also a group homomorphism (Section 1.5) between the abelian group structures on $\mathfrak{g}$ and $F^m$. The kernel of this homomorphism, $\mathrm{Ker}(T)$ consists of $g \in \mathfrak{h}$ such that $\kappa(g, h_i) = 0$ for all $i \in \{1, \ldots, m\}$. In other words, $\mathrm{Ker}(T) = \mathfrak{h}^\perp$. Therefore the quotient group $\mathfrak{g}/\mathfrak{h}^\perp$ is isomorphic to the image of $\mathfrak{g}$ in $F^m$ under the mapping $T$. The reader can easily infer that the group $\mathfrak{g}/\mathfrak{h}^\perp$ is a vector space of dimension equal to $\dim \mathfrak{g} - \dim \mathfrak{h}^\perp$. Since $\mathfrak{g}/\mathfrak{h}^\perp$ is a subspace in $F^m$, it's dimension can be no more than $m$. Thus one has the inequality

$$\dim \mathfrak{g} \leq \dim \mathfrak{h} + \dim \mathfrak{h}^\perp.$$

Since the only common subspace of $\mathfrak{h}$ and $\mathfrak{h}^\perp$ is the null space, $\dim \mathfrak{g} = \dim \mathfrak{h} + \dim \mathfrak{h}^\perp$ and one has $\mathfrak{g} = \mathfrak{h} \oplus \mathfrak{h}^\perp$. Also, the restriction of $\kappa$ to $\mathfrak{h}$ is non-degenerate and $\mathfrak{h}$ is semi-simple. For emphasis, we sum up the arguments of this paragraph as the following:

**Theorem.** *If $\mathfrak{h}$ is an invariant subalgebra of a finitely dimensional semi-simple Lie algebra $\mathfrak{g}$ and $\mathfrak{h}^\perp$ is the orthogonal complement of $\mathfrak{h}$ with respect to the Killing form on $\mathfrak{g}$, then $\mathfrak{g} = \mathfrak{h} \oplus \mathfrak{h}^\perp$. Additionally, $\mathfrak{h}$ is semi-simple and $\mathfrak{h}^\perp$ is an invariant subalgebra.*

It is clear from the above that a semi-simple Lie algebra can have no non-trivial invariant abelian subalgebras. Also note the following

**Definition 9.** A *simple Lie algebra* that has no non-trivial invariant subalgebras is a semi-simple Lie algebra.

It is an obvious consequence of the above theorem that a semi-simple Lie algebra is a direct sum of simple Lie algebras.

## 5.5 Lie Algebra of a Lie Group

**Definition 10.** A Lie group is a group with the structure of an *analytic manifold* in which group multiplication and inversion are analytic functions of their arguments.

The precise concept of an analytic manifold will not be made here. The analytic nature of the group product will be central in subsequent development. The fact that a Lie group is an analytic manifold leads to the consequence that any small neighborhood of the identity of the group has the structure of a vector space. In the example of the circle group, this vector space is one-dimensional and in the exponential representation $\exp(i\omega t)$, it is spanned by the basis $[\omega]$. Generally, the number of parameters needed to span the local vector space near the group identity may be more than one. It is assumed that this number is always finite. Let $G$ be a Lie group whose elements in the vicinity of identity can be parametrized by $n$ real numbers $t^\alpha$[†]. Then $G$ is said to be of dimension $n$. The identity itself is parametrized so that each of $t^\alpha$ is set to zero. For two group elements $g_1$ and $g_2$ parametrized by the sets of real quantities $t_1^\alpha$ and $t_2^\alpha$ respectively, the parameters $t^\alpha$ of their product $g = g_1 g_2$ are analytic functions of $t_1^\alpha$ and $t_2^\alpha$. Suppose

$$t^\alpha = f^\alpha(t_1, t_2),$$

where $t_1$ and $t_2$ represent the complete sets of $n$ parameters of $g_1$ and $g_2$ respectively. Letting $t_1 = 0$ or $t_2 = 0$ to be parameters for the identity element, the condition on $f^\alpha$ obtained is

$$t^\alpha = f^\alpha(t, 0) = f^\alpha(0, t).$$

---

[†]The superscript $\alpha$ should not be confused as an exponent.

With this condition, and the fact that the functions $f^\alpha$ are analytic functions of $t_1$ and $t_2$, a Taylor series for $f^\alpha$ may be written as

$$t^\alpha = f^\alpha(t_1, t_2) = t_1^\alpha + t_2^\alpha + \frac{1}{2}\sum_{\gamma,\delta} f^\alpha_{\gamma\delta} t_1^\gamma t_2^\delta + \cdots \quad (5.5.1)$$

where $f^\alpha_{\gamma\delta}$ are derivatives of $f^\alpha$ at $t_1 = t_2 = 0$. If $\Gamma$ is a unitary representation of $G$ (assuming such a representation exists) and $g$ an element of $G$ close to identity then the representing matrix $\Gamma(g)$ of $g$ is an analytic function of the parameters $t^\alpha$ for $g$.

$$\Gamma(g) = I + i\sum_\alpha t^\alpha X_\alpha - \frac{1}{2}\sum_{\alpha,\beta} t^\alpha t^\beta X_{\alpha\beta} + \cdots \quad (5.5.2)$$

where $X_\alpha$, $X_\alpha X_\beta$ etc are matrices of the same order and do not depend on the value of the real parameters $t^\alpha$. Suppose $g = g_1 g_2$, one has $\Gamma(g_1)\Gamma(g_2) = \Gamma(g)$. The parameters of $g$ are given by Equation 5.5.1. Similar expressions for $\Gamma(g_1)$ and $\Gamma(g_2)$ in combination with $\Gamma(g_1)\Gamma(g_2) = \Gamma(g)$ give

$$\left[I + i\sum_\alpha t_1^\alpha X_\alpha - \frac{1}{2}\sum_{\alpha,\beta} t_1^\alpha t_1^\beta X_{\alpha\beta} + \cdots\right] \times$$

$$\left[I + i\sum_\alpha t_2^\alpha X_\alpha - \frac{1}{2}\sum_{\alpha,\beta} t_2^\alpha t_2^\beta X_{\alpha\beta} + \cdots\right] =$$

$$I + i\sum_\alpha \left[t_1^\alpha + t_2^\alpha + \frac{1}{2}\sum_{\gamma,\delta} f^\alpha_{\gamma,\delta} t_1^\gamma t_2^\delta \cdots\right] X_\alpha -$$

$$\frac{1}{2}\sum_{\alpha,\beta} \left[t_1^\alpha + t_2^\alpha + \frac{1}{2}\sum_{\gamma,\delta} f^\alpha_{\gamma,\delta} t_1^\gamma t_2^\delta \cdots\right] \times$$

$$\left[t_1^\beta + t_2^\beta + \frac{1}{2}\sum_{\gamma,\delta} f^\beta_{\gamma,\delta} t_1^\gamma t_2^\delta \cdots\right] X_{\alpha\beta} + \cdots$$

Expanding the above equation, the reader should observe that the leading terms remaining after cancellations are quadratic in $. For reference, an intermediate step is

$$-\sum_{\alpha,\beta} t_1^\alpha t_2^\beta X_\alpha X_\beta = \frac{i}{2}\sum_{\alpha,\gamma,\delta} t_1^\gamma t_2^\delta f^\alpha_{\gamma,\delta} X_\alpha - \frac{1}{2}\sum_{\alpha,\beta}[t_1^\alpha t_2^\beta + t_2^\alpha t_1^\beta] X_{\alpha\beta}.$$

Here one may notice that the $X_{\alpha\beta} = X_{\beta\alpha}$ since these are coefficients in the Taylor expansion Equation. 5.5.2. After relabelling the indices in the first term in the right hand side of above and using the fact that $t_i^\alpha$ are independent, one has

$$-X_\alpha X_\beta = \frac{i}{2}\sum_\gamma f^\gamma_{\alpha\beta} X_\gamma - X_{\alpha\beta}. \qquad (5.5.3)$$

Interchanging $\alpha$ and $\beta$ in above equation and subtracting gives

$$[X_\alpha, X_\beta] = \frac{i}{2}\sum_\gamma (f^\gamma_{\beta\alpha} - f^\gamma_{\alpha\beta}) X_\gamma, \qquad (5.5.4)$$

where the $[X_\alpha, X_\beta]$ stands for the commutator of $X_\alpha$ and $X_\beta$. The reader will immediately recognize the above equation as the commutation relation in a Lie algebra (Equation 5.3.1) with the structure constants $c^\gamma_{\alpha\beta} = \frac{1}{2}\sum_\gamma (f^\gamma_{\beta\alpha} - f^\gamma_{\alpha\beta})$. The structure constant $c^\gamma_{\alpha\beta}$ is clearly antisymmetric with respect to interchange of indices $\alpha$ and $\beta$. In the other direction, if the structure constant of this Lie algebra are known, then Equation 5.5.3 allows us to calculate $X_{\alpha\beta}$ and consequently, the representing matrix for any group element near the identity up to a precision of second order by use of Equation 5.5.2.

A more general result is that in fact, all the matrices in the infinite series in the right hand side of Equation 5.5.2 are known from the commutations in Equation 5.5.4. This result will not be proven here. An immediate consequence is that knowledge of $X_\alpha$ and its commutations are sufficient to calculate $\Gamma(g)$ for all $g$ in a finite neighborhood of the identity. For this reason all the constituents of $X_\alpha$ are called the *generators* of the group and the vector space spanned by generators is the Lie algebra $\mathfrak{g}$ associated with the group $G$. A Lie group whose Lie algebra is real is called a *real Lie group* and similarly for *complex Lie group* corresponds to Lie algebra which is complex. In the following section, we will discuss examples of Lie groups.

## 5.6 Examples of Lie Groups

Important examples of Lie groups are the *general linear groups* and some of their subgroups. These are Lie groups which can be faithfully represented by matrices of finite size, and so are called *matrix Lie groups*.

The general linear group of degree $n$ is the group of invertible $n \times n$ matrices with the group operation being the usual multiplication of matrices. The general linear group of degree $n$ whose representing matrices have only real entries is denoted by $GL(n, \mathfrak{R})$. In the complex case the notation is $GL(n, \mathfrak{C})$. The group $GL(n, \mathfrak{R})$ is itself a subgroup of $GL(n, \mathfrak{C})$. Regarded as Lie groups on their own, there is an important difference between $GL(n, \mathfrak{R})$ and $GL(n, \mathfrak{C})$. $GL(n, \mathfrak{R})$ consists of invertible matrices whose determinants are non vanishing. Therefore these determinants can be either positive valued or negative valued.

For $\mathbf{M_1}$ and $\mathbf{M_2}$ in $GL(n, \mathfrak{R})$ such that $\det \mathbf{M_1} > 0$ and $\det \mathbf{M_2} < 0$, there cannot then exist a continuous real parameter smoothly connecting $\mathbf{M_1}$ to $\mathbf{M_2}$. On the otherhand, $GL(n, \mathfrak{C})$ does not suffer from this discontinuity as it is always possible to find a path in the complex plane. $GL(n, \mathfrak{C})$ is therefore an example of a *connected Lie group*. $GL(n, \Re)$ can be decomposed into two components. One component consists of matrices with positive determinant while the other of matrices with negative determinant. Each component is connected in itself but disconnected from the other. The identity matrix of $GL(n, \Re)$ lies in the positive component.

*Special linear groups* are subgroups of general linear groups. The corresponding notations are $SL(n, \Re)$ and $SL(n, \mathfrak{C})$. These groups consist of matrices whose determinants equal unity. As before, $SL(n, \Re)$ is a subgroup of $SL(n, \mathfrak{C})$. Both the groups are connected Lie groups.

The *orthogonal group* $O(n)$ is a subgroup of $GL(n, \Re)$. The defining condition for a matrix $\mathbf{M} = [m_{ij}]_{n \times n}$ to be in $O(n)$ is that

$$\mathbf{M}^T \mathbf{M} = \mathbf{M}\mathbf{M}^T = \mathbf{I}. \tag{5.6.1}$$

Upon explicitly writing out the product $\mathbf{M}^T \mathbf{M}$, the above condition reduces to $\sum_{i=1}^{n} m_{ij} m_{ik} = \delta_{jk}$. In other words, the columns of any matrix in $O(n)$ are orthonormal. The same can be inferred about the rows. From the defining Equation 5.6.1 it also follows that the determinant of the matrices in $O(n)$ can be either $+1$ or $-1$. The subgroup of $O(n)$ consisting of matrices whose determinants equal unity is called the *special orthogonal group* $SO(n)$. The group $SO(n)$ essentially consists of transformations that preserve handedness (rotations), while in $O(n)$ there are also improper transformations like reflections which do not preserve handedness.

The *generalized orthogonal group* $O(m, n)$ is another subgroup of $GL(m + n, \Re)$. Define $\mathbf{g}$ to be a $(m + n) \times (m + n)$ matrix in which the the first $m$ entries in the principal diagonal are $+1$, the last $n$ entries in the principal diagonal are $-1$ while all the other entries are 0. A real matrix $\mathbf{M}$ is in $O(m, n)$ if and only if

$$\mathbf{M}^T \mathbf{g} \mathbf{M} = \mathbf{M} \mathbf{g} \mathbf{M}^T = \mathbf{g}. \tag{5.6.2}$$

It is obvious from the above definition that $\det \mathbf{M} = \pm 1$. The subgroup of $O(m, n)$ which contains only matrices of unit determinant is the *generalized special orthogonal group* $SO(m, n)$. In fact, the *Lorentz group* is the group $SO(1,3)$. As this group is the symmetry group of physical systems respecting the laws of special theory of relativity, we will elaborate the representations of Lorentz group in the next chapter.

The *symplectic group* $Sp(2n, \Re)$ is another subgroup of $SL(2n, \Re)$. The defining equation of a symplectic matrix $M_{2n \times 2n} \in Sp(2n, \Re)$ is

$$\mathbf{M}^T \mathbf{J} \mathbf{M} = \mathbf{J},$$

where

$$J = \begin{bmatrix} 0_{n \times n} & I_{n \times n} \\ -I_{n \times n} & 0_{n \times n} \end{bmatrix}.$$

Symplectic groups find applications in the transformation properties of the position and velocity variables describing classical systems. Now turning to $GL(n, \mathbb{C})$, the subgroup consisting of such matrices **M** for which

$$\mathbf{M}^\dagger \mathbf{M} = \mathbf{M}\mathbf{M}^\dagger = \mathbf{I} \tag{5.6.3}$$

is called the *unitary group* $U(n)$. The circle group $U(1)$ was the first example of this type. It is obvious from the above definition that determinants of matrices in $U(n)$ are unimodular complex numbers. The subgroup of $U(n)$ containing matrices whose determinants equal unity is call the *special unitary group* $SU(n)$. The group $SU(2)$ plays an important role in description of particle spin in quantum mechanics. Hence their applications in quantum mechanics and particle physics will be discussed in the following chapter.

We will now discuss the classification of semi-simple Lie algebras known in the literature as *Cartan classification*.

## 5.7 Compact Simple Lie Algebras

In Section 5.5, we came across the fact that the Lie algebra of a Lie group can reveal the structure of the group in a finite neighborhood of the identity element. Here, we will focus on the broad classification of simple Lie algebras.

The works of Killing and Cartan led to the classification of simple Lie algebras $\mathfrak{g}$ to be of four infinite types and five exceptional types. We will now present the definitions that govern operations followed by the $n$ basis elements of Lie algebra $\mathfrak{g}$ without proofs.

Let $\ell$ independent elements $\{H_i\} \in \mathfrak{g}$ obey abelian subalgebra:

$$[H_i, H_j] = 0; \quad i, j = 1, \ldots \ell$$

which is known in the literature as *Cartan subalgebra*. The number of commuting elements ($\ell$) of the algebra $\mathfrak{g}$ is called the *rank* of the algebra.

The remaining $n - \ell$ elements $E_{\pm\alpha} \in \mathfrak{g}$ are even in number. Note that the subscript $\pm\alpha$ are $\ell$-component vectors which are referred to as positive and negative *root vectors* of the algebra. Hence, there will be $n - \ell$ non-zero root vectors associated with the Lie algebra $\mathfrak{g}$. The algebra involving $E_{\pm\alpha}$ will yield

$$[E_\alpha, E_{-\alpha}] = \sum_{i=1}^{\ell} \alpha_i H_i,$$

where $\alpha_i$ refers to the i-th component of the root vector $\alpha$. If $\alpha$ and $\beta$ are two different root vectors of $\mathfrak{g}$ then

$$[E_\alpha, E_\beta] = \mathcal{N}_{\alpha,\beta} E_{\alpha+\beta},$$

where $\mathcal{N}_{\alpha,\beta} = -\mathcal{N}_{\beta,\alpha} \neq 0$ if and only if $\alpha + \beta$ is also a root vector of $\mathfrak{g}$. Further, for complete definition of the algebra $\mathfrak{g}$, we need the following commutator as well:

$$[H_i, E_\alpha] = \alpha_i E_\alpha;\ [H_i, E_{-\alpha}] = -\alpha_i E_{-\alpha}.$$

Note that the above commutator bracket resembles ladder operators and $J_3$ of $\mathfrak{su}(2)$ algebra (5.3.10). As $\mathfrak{su}(2)$ algebra has only one diagonal generator, the rank of the $\mathfrak{su}(2)$ algebra is $\ell = 1$ leading to one-component root vectors (numbers). The comparison of the formal $\mathfrak{g}$ algebra with $\mathfrak{su}(2)$ implies that there are two non-zero roots $\pm \alpha = \pm 1$ and hence the corresponding generators $E_{\pm 1} \equiv J_\pm$.

For a general Lie algebra $\mathfrak{g}$, we can construct many $\mathfrak{su}(2)$ subalgebras in the following way:

$$J_\pm^{(\alpha)} = |\alpha|^{-1} E_{\pm\alpha};\ J_3^{(\alpha)} = |\alpha|^{-2} \alpha.H.$$

Similar to the $\mathfrak{su}(2)$ eigenbasis $|jm\rangle$ of $J_3$ with half integer values $m$, we have eigenbasis $|\Lambda, \mu\rangle \in V$ of $J_3^{(\alpha)}$ where $\Lambda$ is the highest weight vector and $\mu$ denotes $\ell$ component weight vector with the following properties:

$$J_3^{(\alpha)}|\Lambda,\mu\rangle = \frac{\alpha.\mu}{|\alpha|^2}|\Lambda,\mu\rangle\ ;\ J_\pm^{(\alpha)}|\Lambda,\mu\rangle \propto |\Lambda,\mu \pm \alpha\rangle$$

where the eigenvalues of $J_3^{(\alpha)}$ are half integers. Hence,

$$\frac{2\alpha.\mu}{|\alpha|^2} = \text{integer}.$$

This integer can be determined by finding $p, q$ such that

$$(J_+^{(\alpha)})^{p+1}|\Lambda,\mu\rangle = 0\ ;\ (J_-^{(\alpha)})^{q+1}|\Lambda,\mu\rangle = 0, \tag{5.7.1}$$

implying that the eigenvalue of $J_3^{(\alpha)}$ is maximum $(+j)$ for state $|\Lambda, \mu + p\alpha\rangle$ and minimum $(-j)$ for state $|\Lambda, \mu - q\alpha\rangle$. That is,

$$\frac{\alpha.(\mu + p\alpha)}{|\alpha|^2} = j = -\frac{\alpha.(\mu - q\alpha)}{|\alpha|^2},$$

which implies

$$\frac{2\alpha.\mu}{|\alpha|^2} = (q - p). \tag{5.7.2}$$

Comparing with the $\mathfrak{su}(2)$ ladder operations, we can say highest weight state $|\Lambda, \Lambda\rangle$ is defined as

$$J_+^{(\alpha)}|\Lambda, \mu = \Lambda\rangle = 0 \,\forall\, \alpha.$$

Suppose we choose the vector space $V = \mathfrak{g}$ (adjoint representation), then the eigenbasis of the $J_3^{(\alpha)}$ can be represented by $\{|H_i\rangle\}$, $\{|E_{\pm\beta}\rangle\}$. In fact, for adjoint representation, the weight vector of states are root vectors. Applying $J_3^{(\alpha)}$, $J_3^{(\beta)}$ on such states will give the following eigenvalue equations:

$$J_3^{(\alpha)}|\Lambda, E_{+\beta}\rangle = \frac{\alpha.\beta}{|\alpha|^2}|\Lambda, E_{+\beta}\rangle \,;\, J_3^{(\beta)}|\Lambda, E_{+\alpha}\rangle = \frac{\alpha.\beta}{|\beta|^2}|\Lambda, E_{+\alpha}\rangle.$$

Using the two equations (Equation 5.7.1 and 5.7.2) for the above states, we can infer:

$$\frac{2\alpha.\beta}{|\alpha|^2} = (q-p), \quad \frac{2\beta.\alpha}{|\beta|^2} = (q'-p'),$$

suggesting that the angle between two root vectors must obey:

$$\cos^2\theta_{\alpha,\beta} = \frac{(q-p)(q'-p')}{4},$$

where $p$, $q$, $p'$, $q'$ are integers. This forces the non-trivial angle between two root vectors to assume values of $90^0, 60^0$ or $120^0, 45^0$ or $135^0, 30^0$ or $150^0$.

If the rank of the Lie algebra $\mathfrak{g}$ is $\ell$, then there are $\ell$ *simple roots*. The remaining positive and negative roots can be obtained from linear combination of the simple roots. There is a neat concise way of classifying simple Lie algebras $\mathfrak{g}$ using connected graphs called Dynkin diagrams. Suppose we denote the $\ell$ simple root vectors by $\ell$ filled circles. We connect these $\ell$ filled circles by single bond or double bond or triple bond depending on the angle between any two simple roots $\alpha$, $\beta$. In fact,

$$4\cos^2\theta_{\alpha,\beta} = m,$$

where $m$ is the number of bonds connecting the simple roots $\alpha$, $\beta$. For instance, the Lie algebra $A_\ell \equiv \mathfrak{su}(\ell+1)$ has $\ell$ simple roots of equal length ($\alpha^{(1)}, \alpha^{(2)}, \ldots \alpha^{(\ell)}$) and has a single bond connecting adjacent simple roots as shown in Figure 5.7.1. That is,

$$4\cos^2\theta_{\alpha^{(i)}, \alpha^{(i\pm 1)}} = 1.$$

Besides the $A_\ell$ infinite series Dynkin diagram depicting unitary algebras $\mathfrak{su}(\ell+1)$ of arbitrary ranks, we have other connected graphs corresponding to orthogonal algebras $so(n)$ which belong to infinite series Dynkin diagrams $B_{\ell \geq 3}$ for $n = 2\ell+1$

and $D_{\ell \geq 4}$ for $n = 2\ell$ as shown in Figure 5.7.1. Note that the last two simple roots in $B_\ell$ are connected by double bond and the orientation of the arrow indicates that the length of the root vectors obey $|\alpha^{(i)}| > |\alpha^{(\ell)}|$ for all $i < \ell$. The Dynkin diagram of the symplectic algebra $C_{\ell \geq 2} \equiv sp(2\ell)$ has a reversed orientation implying $|\alpha^{(i)}| < |\alpha^{(\ell)}|$ for all $i < \ell$. Besides these connected graphs of arbitrary ranks (see Figure 5.7.1(a)), we can have finite rank Dynkin diagrams which are known in the literature as five exceptional Lie algebras (as shown in Figure 5.7.1(b)). Basically, these four infinite series and five exceptional series are the only allowed connected graphs for linearly independent simple roots which can represent simple Lie algebras $\mathfrak{g}$.

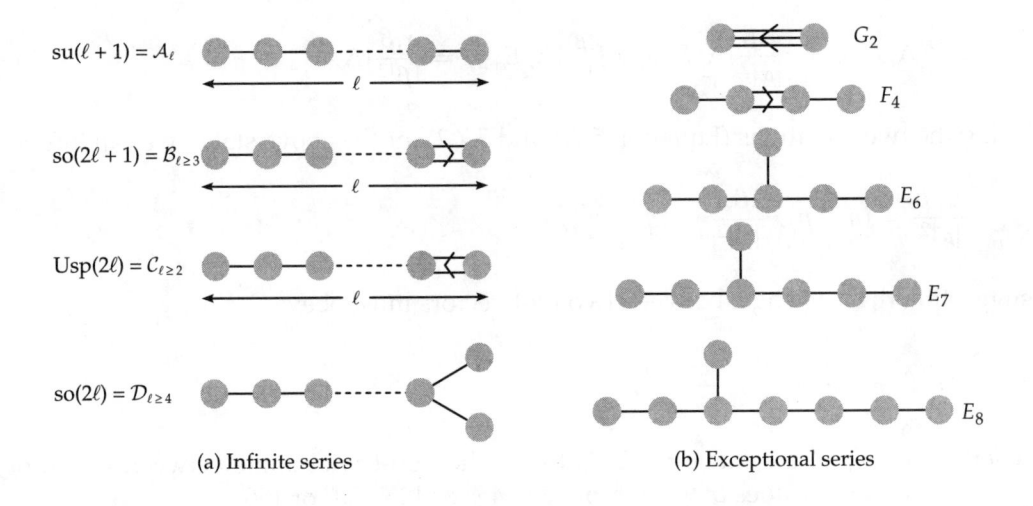

**Figure 5.7.1** Dynkin Diagrams

With the formal definition of simple Lie algebras where we introduced root and weight vectors, we will now extensively discuss $A_{\ell=2} \equiv \mathfrak{su}(3)$ algebra.

### 5.7.1 $\mathfrak{su}(3)$ algebra

Just like the $2 \times 2$ $\sigma$ matrices are the fundamental representation of $\mathfrak{su}(2)$ acting on a two-dimensional complex vector space $V$, we have $3 \times 3$ Gell-Mann matrices $\{\lambda_a\}$'s which are referred to as the defining representation of $\mathfrak{su}(3)$ acting on three-dimensional complex vector space $V$. These Gell-Mann matrices are Hermitian and traceless. Hence, there will be 8 $\lambda_a$'s denoting the basis elements of $\mathfrak{su}(3)$ algebra because hermiticity and traceless conditions reduce the nine complex entries of these $3 \times 3$ matrices to eight independent real entries. The explicit form of the Gell-Mann matrices are

$$\lambda_1 = \begin{pmatrix} 0 & 1 & 0 \\ 1 & 0 & 0 \\ 0 & 0 & 0 \end{pmatrix}, \lambda_2 = \begin{pmatrix} 0 & -i & 0 \\ i & 0 & 0 \\ 0 & 0 & 0 \end{pmatrix}, \lambda_3 = \begin{pmatrix} 1 & 0 & 0 \\ 0 & -1 & 0 \\ 0 & 0 & 0 \end{pmatrix}$$

# Lie Groups and Lie Algebras

$$\lambda_4 = \begin{pmatrix} 0 & 0 & 1 \\ 0 & 0 & 0 \\ 1 & 0 & 0 \end{pmatrix}, \lambda_5 = \begin{pmatrix} 0 & 0 & -i \\ 0 & 0 & 0 \\ i & 0 & 0 \end{pmatrix}, \lambda_6 = \begin{pmatrix} 0 & 0 & 0 \\ 0 & 0 & 1 \\ 0 & 1 & 0 \end{pmatrix}$$

$$\lambda_7 = \begin{pmatrix} 0 & 0 & 0 \\ 0 & 0 & -i \\ 0 & i & 0 \end{pmatrix}, \lambda_8 = \frac{1}{\sqrt{3}} \begin{pmatrix} 1 & 0 & 0 \\ 0 & 1 & 0 \\ 0 & 0 & -2 \end{pmatrix}.$$

Note that there are two diagonal matrices $\lambda_3$, $\lambda_8$ which constitute the Cartan subalgebra of rank $\ell = 2$. Hence the three basis states

$$|\Lambda\mu_1\rangle = \begin{pmatrix} 1 \\ 0 \\ 0 \end{pmatrix}, |\Lambda\mu_2\rangle = \begin{pmatrix} 0 \\ 1 \\ 0 \end{pmatrix}, |\Lambda\mu_3\rangle = \begin{pmatrix} 0 \\ 0 \\ 1 \end{pmatrix}.$$

of the fundamental three-dimensional complex vector space $V$ are simultaneous eigenstates of both $H_1 = \lambda_3/2$ and $H_2 = \lambda_8/2$. We have indicated both the eigenvalues in the two-dimensional $H_1 - H_2$ diagram (Figure 5.7.2) for the three basis states. Hence, we can equivalently represent the basis state as two-component weight vectors:

$$|\Lambda\mu_1\rangle \equiv (1/2, \sqrt{3}/6); |\Lambda\mu_2\rangle \equiv (-1/2, \sqrt{3}/6); |\Lambda\mu_3\rangle \equiv (0, -\sqrt{3}/3).$$

The weight vectors are called negative weight vectors if the first non-zero component is negative. In the above set of weight vectors, $\mu_1$ is positive weight vector whereas $\mu_2, \mu_3$ are negative weight vectors.

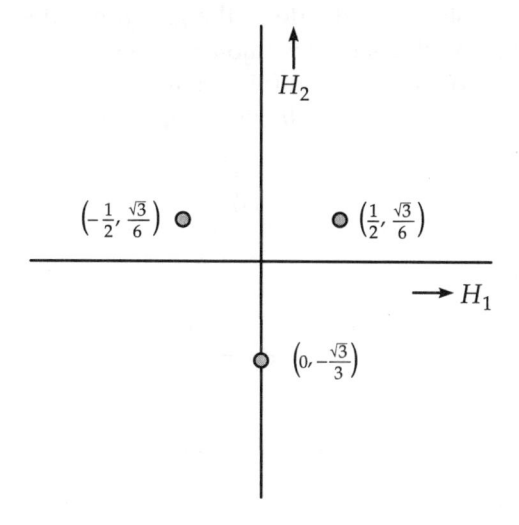

**Figure 5.7.2** su(3) Weight Diagram for Defining Representation

The raising and lowering operators will take one basis state to another basis state. The explicit form of such operators will involve the remaining six Gell-Mann matrices in the following way:

$$E_{\pm\alpha^{(3)}} = \frac{1}{2\sqrt{2}}(\lambda_1 \pm i\lambda_2) \,;\; E_{\pm\alpha^{(1)}} = \frac{1}{2\sqrt{2}}(\lambda_4 \pm i\lambda_5);$$

$$E_{\mp\alpha^{(2)}} = \frac{1}{2\sqrt{2}}(\lambda_6 \pm i\lambda_7),$$

where

$$E_{-\alpha^{(3)}}|\Lambda,\mu_1\rangle \propto |\Lambda,\mu_1 - \alpha^{(3)}\rangle = |\Lambda,\mu_2\rangle \,;\; E_{+\alpha^{(i)}}|\Lambda,\mu_1\rangle = |\Lambda,\mu_1 + \alpha^{(i)}\rangle = 0 \,\forall\, i, \quad (5.7.3)$$

implying that the weight vector $|\Lambda,\mu_1\rangle$ is the *highest weight state* (that is., $\Lambda = \mu_1$ and the root vector

$$\alpha^{(3)} = \mu_1 - \mu_2 = (1,0).$$

In the similar fashion, you can check that

$$\alpha^{(1)} = \mu_1 - \mu_3 = (1/2, \sqrt{3}/2), \; \alpha^{(2)} = \mu_3 - \mu_2 = (1/2, -\sqrt{3}/2). \quad (5.7.4)$$

Note that $\alpha^{(3)} = (1,0) = \alpha^{(1)} + \alpha^{(2)}$ indicating that the two simple roots of the $\mathfrak{su}(3) \equiv A_2$ algebra are $\alpha^{(1)}$, $\alpha^{(2)}$ whereas $\alpha^{(3)}$ is not a simple root vector. These non-zero root vectors can be plotted in the $H_1 - H_2$ graph (see Figure 5.7.3 where the right side of the graph has positive roots and left side of the graph has negative roots) which are the non-zero weight vector states of the adjoint representation which also represent the raising and lowering operators of $\mathfrak{su}(3)$ algebra. The origin of the graph are the null root vectors denoting the Cartan subalgebra operators $H_1$, $H_2$.

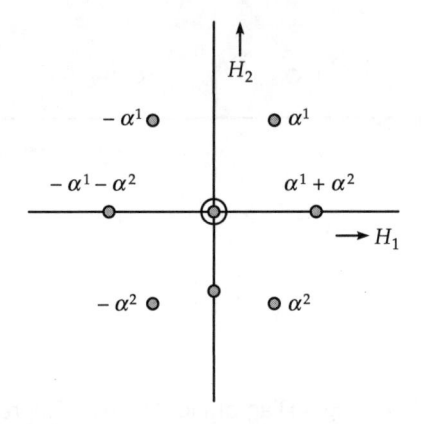

**Figure 5.7.3**    $\mathfrak{su}(3)$ Root Diagram

## 5.7.2 Cartan matrix

For a Lie algebra of rank $\ell$, there will be $\ell$ simple roots. One can construct the Cartan matrix

$$A_{ij} = \frac{2\alpha^{(i)}.\alpha^{(j)}}{|\alpha^{(i)}|^2}$$

from the Dynkin diagram. For $A_2 \equiv su(3)$, the Cartan matrix elements using the simple root vectors (5.7.4):

$$\begin{pmatrix} 2 & -1 \\ -1 & 2 \end{pmatrix}.$$

## 5.7.3 Fundamental weights

We will have $\ell$ fundamental highest weight vectors $\mu^{(i)}$'s such that

$$\frac{2\alpha^{(i)}.\mu^{(j)}}{|\alpha^{(i)}|^2} = \delta_{ij}. \tag{5.7.5}$$

**Example 48.** For $su(3)$ algebra, determine the two fundamental highest weight vectors.

Let us take the 2-component fundamental highest weight vector as $(u, v)$, Using the orthogonality property (5.7.5) with simple roots (5.7.4), we get $u = \pm\sqrt{3}v$. The normalized form of the two fundamental highest weight vectors are

$$\mu^{(1)} = (1/2, \sqrt{3}/6) \, ; \, \mu^{(2)} = (1/2, -\sqrt{3}/6). \qquad \square$$

Recall $\mu^{(1)}$ (5.7.3) is the highest weight vector of the defining representation which we discussed using Gell-Mann matrices. For arbitrary representation of $su(3)$, the highest weight vector will be

$$\Lambda \equiv \mu = a\mu^{(1)} + b\mu^{(2)}$$

where $a, b$ are positive integers. For example, the highest weight vector of the adjoint representation is $\alpha^{(1)} + \alpha^{(2)} = \mu^{(1)} + \mu^{(2)} = (1, 0)$.

Armed with the formal aspects of Lie groups and Lie algebras discussed in this chapter, we will focus on their applications in the following chapter. Also, the tensor product of irreducible representations of $su(2)$ and $su(3)$ algebra will be discussed in detail.

## Exercises

1. We have discussed that the group $U(2)$ has four independent parameters. Suppose we impose that the determinant of these unitary matrices to be $+1$, then we call the group as $SU(2)$. What will be the number of independent parameters for $SU(2)$ group.

2. Determine the number of independent parameters (equal to number of generators) for a general $U(N)$ group and $SU(N)$ group.

3. For a Lie algebra $\mathfrak{g}$ in which the structure constants are $c_{st}^k$ as in Equation 5.3.1, show that

$$\sum_{k=1}^{n} [c_{sk}^l c_{tu}^k + c_{tk}^l c_{us}^k + c_{uk}^l c_{st}^k] = 0$$

[Hint: Jacobi identity]

4. Let $\mathfrak{g}$ be a semi-simple Lie algebra and $\mathfrak{h}$ an invariant subalgebra in $\mathfrak{g}$. Show that $(\mathfrak{h}^\perp)^\perp = \mathfrak{h}$.

5. For vectors x and y in $\mathfrak{R}^n$, define the product of x and y as

$$x.y = \sum_{i=1}^{n} x_i y_i.$$

Show that under a transformation $\mathbf{M}$ in $O(n)$, the value of the product is invariant, i.e.,

$$x.y = Mx.My.$$

6. For vectors x and y in $\mathfrak{R}^{m+n}$, define the product of x and y as

$$x.y = \sum_{i=1}^{m} x_i y_i - \sum_{j=1}^{n} x_{m+j} y_{m+j}.$$

Show that under a transformation $\mathbf{M}$ in $O(m, n)$, the value of the product is invariant.

7. For vectors x and y in an $n$-dimensional complex vector space, define the product of x and y as

$$xy = \sum_{i=1}^{n} \overline{x_i} y_i$$

where $\overline{x_i}$ represents the complex conjugate of $x_i$. Show that under the transformations in $U(n)$, the value of the product is invariant.

8. Given a Lie algebra $\mathfrak{g}$ having two simple roots

$$\alpha^{(1)} = (0, 1), \quad \alpha^{(2)} = (\sqrt{3}/2, -3/2),$$

work out the Cartan matrix and the corresponding Dynkin diagram. Also determine their fundamental highest weight vectors.

# 6

# Further Applications

With the formal aspects of Lie algebra and Lie groups discussed in the previous chapter, it is now possible to apply the power of continuous symmetry to physics of many simple and complex systems. This chapter will focus upon the applications of Lie algebra and groups to physical systems possessing orthogonal group or unitary group symmetry. For completeness, note that the symplectic group $sp(2n)$ is nothing but the canonical transformation of $2n$ phase space variables (position and velocity variables). Readers are advised to look up canonical transformations in any classical mechanics textbooks such as Goldstein for further appreciation of the natural emergence of symplectic group symmetry in classical systems.

Many classical systems in three-dimensional space respect spherical symmetry. Such systems remain unchanged under proper rotations. Noether's theorem states that 'associated with any continuous group symmetry, there is always a constant of motion'. In Section 6.1, we will briefly discuss classical systems possessing translational symmetry and show that the constant of motion is linear momentum **p**.

We have extensively discussed, in the previous chapter, the rotation group $SO(3)$ and its Lie algebra $so(3)$ whose generators are the three components of orbital angular momentum **L**. Thus the generators of the group are nothing but constants of motion in the classical system respecting rotational symmetry. The rotational invariance of quantum mechanical systems leads to a natural emergence of group $SU(2)$ extensively presented in the previous chapter. Particularly, the Lie algebra generators are the components of the total angular momentum operator $\hat{\mathbf{J}} = \hat{\mathbf{L}} + \hat{\mathbf{S}}$ incorporating the intrinsic spin $\hat{\mathbf{S}}$ in accordance with the famous Stern–Gerlach experiment. Any basic course on the quantum mechanics of rotational invariant systems is based on the theory of angular momentum algebra which is isomorphic to $su(2)$.

In the previous chapter, we have discussed the actions of Lie algebra generators $E_{\pm\alpha}$, associated with a root vector $\alpha$, on the highest weight vector $\Lambda$ leading to weight vectors $\Lambda \pm \alpha$. Through $\mathfrak{su}(2)$ as an example (Chapter 5, Section 5.3.1), we showed that angular momentum corresponds to the highest weights in the Lie group context and the the range of z–components of angular momentum are the weights. In Section 6.2, we will construct the tensor product of two or more angular momentum states using the Young tableau approach. In quantum mechanics literature, they are known as *addition of angular momentum*. In Subsection 6.2.1, we have a small digression on Young Tableau approach for both symmetry group of degree $n$ and $SU(N)$ group. Then, we will discuss *selection rules* for the quantum mechanical transition from initial state to final state due to interactions. The important theorem known as *Wigner–Eckart theorem* will be dealt with proof. Through examples, readers will come to appreciate the power and elegance of $SU(2)$ group theory tools in validating experimentally observed allowed and forbidden transitions in quantum systems possessing rotational symmetry. We also remind the readers to compare the selection rules due to discrete symmetry discussed in Chapter 4 to the Wigner–Eckart theorem rules discussed here.

This theorem is applicable to many other systems possesing $SU(2)$ symmetry. For instance, theoretical validation of (i) almost same mass $m_p \approx m_n$ of proton and neutron (ii) experimentally observed strong interaction processes (scattering and decay) involving such elementary particles (broadly classified as *baryons* and *mesons*) are deduced invoking $SU(2)$ Lie group symmetry. For baryons and mesons, $SU(2)$ represents rotational symmetry in an abstract internal space called *isospin* space whose algebra is same as $\mathfrak{su}(2)$. We discuss these features in elementary particle physics in Section 6.3.

Interestingly, these elementary particles with same angular momentum **J** but with different isospin **I** and charge $Q$ can be plotted on a two-dimensional diagram which appears identical to the $\mathfrak{su}(3)$ weight diagrams discussed in the previous chapter. Historically, Gell-Mann introduced a simple model called the *quark model* to theoretically predict the baryons and mesons observed in the laboratory. According to the quark model, particles like protons, neutrons and other baryons must be composed of three quarks. Assuming that there are three flavors of quarks: u, d, s (up, down and strange), Gell-Mann applied the unitary group $SU(3)$ and tensor product of quarks to account for baryons observed in experiments. Interestingly, he predicted that the $SU(3)$ symmetry requires the presence of a baryon $\Omega^-$ which was detected experimentally three years later. This is one of the instances where an abstract theory was validated by the experiment after some years. We discuss the quark model in Section 6.4. Similar to the symmetry breaking discussed in the discrete group context, we will briefly present symmetry breaking in the context of continuous groups. The $SU(3)$ quark model can be generalized to $SU(N)$ assuming quarks come in N flavors using Young diagrams and their tensor product discussed in Subsection 6.2.1.

Besides compact groups, there exist non-compact groups whose Lie algebra can be systematically deduced. In Section 6.5, we will elaborate on one such group called Lorentz group : $SO(3,1)$. One of the reasons for focussing on this Lorentz group is because it is the symmetry possessed by relativistic particles in the $3+1$ dimensional

Minkowski space-time. We will briefly discuss Poincare algebra and Poincare groups which serve as a underlying symmetry of many physical systems. Many systems near criticality (near phase transition point) are invariant under scale transformation. The group symmetry possessed by such systems are known as *conformal groups*. We will give an overview of the conformal group symmetry and their generators. Finally, the readers can see through the exercise problem, the resemblence between the Lorentz algebra $\mathfrak{so}(3,1)$ and $\mathfrak{so}(4)$ associated with rotations in four-dimensional space.

Even though four-dimensional space is not physical, we will show in Section 6.6 that the abstract group $SO(4)$ has an elegant way of reproducing familiar hydrogen atom energy levels without solving the Schrödinger equation. Besides angular momentum **L**, the hydrogen atom has another constant of motion called Runge–Lenz vector **M**. The abstract group $SO(4)$ generators and their algebra turns out to be isomorphic to the algebra involving **L** and **M**. Unlike angular momentum **L** which can be attributed to geometrical transformation of rotations, **M** has no such geometrical interpretation. Such a symmetry which has no geometrical interpretation is known as *dynamical symmetry*.

## 6.1 Continuous Symmetry and Constant of Motion

We will focus on the familiar translation symmetry to formally understand the corresponding constant of motion (Noether theorem). Further, we will also examine representations of these translation group elements and deduce the generators of such transformation. This detailed description will show that the constants of motion act as generators of the corresponding group transformation.

### 6.1.1 Translational symmetry in three-dimensional space

At the start, we will take the simplest symmetry of a system under the space translation operation. Consider a free particle moving in our three-dimensional space. The classical Hamiltonian $H = p^2/(2m)$ is unchanged under translation $\mathbf{x} \to \mathbf{x} + \mathbf{a}$ for any constant shift vector **a**. For such a translationally invariant classical system, it is easy to show that the Hamilton's equation of motion involving Poisson bracket $\{,\}$ (see *Classical Mechanics* by Goldstein) is

$$\frac{d\mathbf{p}}{dt} = \{H, \mathbf{p}\} = 0,$$

which implies linear momentum **p** is a constant of motion for translationally invariant systems.

In quantum mechanics, the free particle wavefunction $\psi(\mathbf{x})$ under translation operation $\hat{T}(\mathbf{a})$ transforms as follows:

$$\hat{T}(\mathbf{a})\psi(\mathbf{x}) = \psi'(\mathbf{x}),$$

where wavefunctions obey $\psi(\mathbf{x}) = \psi'(\mathbf{x} + \mathbf{a})$ for translation invariant systems. This implies $\psi'(\mathbf{x}) = \psi(\mathbf{x} - \mathbf{a})$. Substituting this functional form in the above equation and writing as a Taylor series expansion, we can deduce the form of the operator from

$$\hat{T}(\mathbf{a})\psi(\mathbf{x}) = \exp(-\mathbf{a}.\nabla)\psi(\mathbf{x}).$$

The differential operator is related to the well-known momentum operator, $\hat{\mathbf{p}} = -i\hbar\nabla$, leading to the explicit form of the translation operator as

$$\hat{T}(\mathbf{a}) = \exp\left(\frac{-i\mathbf{a}.\hat{\mathbf{p}}}{\hbar}\right).$$

For an infinitesimal translation (**a** small), for the above operator the expansion up to $\mathcal{O}(a)$ will be

$$\hat{T}(a) \approx \mathbb{I} - i\mathbf{a}.\hat{\mathbf{p}}/\hbar + \mathcal{O}(a^2).$$

In fact, as discussed in the previous chapter on Lie groups, the deviation from identity $\mathbb{I}$ operator generates the infinitesimal translation operation. Thus for spatial translations, linear momentum $\hat{\mathbf{p}}$ is the generator and $\mathbf{a} \in \mathbb{R}^3$ is the parameter in three-dimensional space. We have explicitly shown that the constant of motion is indeed the linear momentum generating translational symmetry. This methodology can be applied to time translations as well where the generator turns out to be the Hamiltonian $H$ and the corresponding time translation parameter $\tau \in \mathbb{R}$ belongs to a real line. We have already seen rotational symmetry groups $SO(3)$ and their Lie algebra $\mathfrak{so}(3)$ whose generators are the orbital angular momentum **L** in the three-dimensional space. For completeness, rotational symmetry and corresponding constant of motion is presented through the following example.

**Example 49.** For a rotationally invariant system described by the Hamiltonian

$$H = \frac{p^2}{2m} + \frac{k}{r^2},$$

where $r$ is the radial coordinate and constant $k$ is a dimensionful constant. Show that $d\mathbf{L}/dt = 0$, confirming that the generators **L** are constants of motion of such rotational invariant systems.

The Poisson bracket

$$\{H, L_i\} = \sum_j \frac{\partial H}{\partial x_j}\frac{\partial L_i}{\partial p_j} - \frac{\partial H}{\partial p_j}\frac{\partial L_i}{\partial x_j} = \sum_{j,l} \epsilon_{ilj}\left(\frac{k}{r^3}x_j x_l - \frac{1}{m}p_j p_l\right) = 0.$$

Hence the Poisson equation of motion for $d\mathbf{L}/dt$ is zero indicating that the angular momentum is a constant of motion for rotational invariant systems. □

In the context of discrete groups, the tensor product of two or more polar vectors and their decomposition as binary basis, tertiary basis and so on, were dealt with using character tables and projection operators. Interestingly, there is an equivalent systematic procedure to deal with the tensor product of vectors and higher rank tensors, leading to irreducible tensors of $SO(3)$. As the algebra $\mathfrak{so}(3)$ is same as $\mathfrak{su}(2)$ algebra, the construction of a tensor product in the $SU(2)$ context turns out to be applicable to $SO(3)$ as well. Hence in the following section, we will elaborate on this tensor product aspect using the irreducible angular momentum states $|j, m\rangle \in V^j$ (vector space whose dimension is $d_j = 2j+1$ as $m \in [-j, -j+1, \ldots j]$) and the action of $\mathfrak{su}(2)$ generators $J_\pm, J_z$ discussed in the previous chapter. This is the main core of the addition of angular momentum and the construction of irreducible states in quantum mechanics.

## 6.2 Tensor Product Rule for $SU(2)$ Irreducible Representations

Let us take the tensor product of two vector spaces $V^{j_1} \otimes V^{j_2}$ whose decomposition into irreducible components can be formally written as

$$V^{j_1} \otimes V^{j_2} = \sum_j N^j_{j_1, j_2} V^j, \qquad (6.2.1)$$

where it can be shown that $N^j_{j_1, j_2} = 1$ when $|j_1 - j_2| \leq j \leq j_1 + j_2$ and $N^j_{j_1, j_2} = 0$ for the other $j$'s. Basically, the proof for the allowed range of $j$ and minimum value of $j = j_{\min}$ is based on taking maximum $j = j_{\max} = j_1 + j_2$ and constraining the dimension of the vector space to obey

$$d_{j_1} \cdot d_{j_2} = \sum_{j=j_{\min}}^{j_{\max}} N^j_{j_1, j_2} d_j.$$

There is an equivalent elegant Young tableau approach of obtaining such a tensor product decomposition. We present this diagrammatic method where we can directly determine $d_j$ as well as visualize decomposition.

### 6.2.1 Digression on the Young tableau approach

**Symmetric Group $\mathfrak{S}(n)$**

Recall the definition of the Young tableau diagram which we reviewed in the context of cycle structure of the elements of the symmetric group $\mathfrak{S}(n)$. We can also use the set of integers $\lambda_1 \geq \lambda_2 \ldots$, associated with a Young diagram $Y$ to denote an irreducible representation of $\mathfrak{S}(n)$. Here $\lambda_1$ denotes the number of boxes in first row, $\lambda_2$ being the number of boxes in second row and so on such that $\sum_i \lambda_i = n$. Note that $n$ entries in these boxes must be distinct for a permutation group. The diagram implies totally symmetric property along any row and totally antisymmetric along any column.

For instance, an irreducible representation $Y \in \mathfrak{S}(5)$ denoted by

$$\begin{array}{|c|c|c|} \hline 1 & 2 & 3 \\ \hline 4 & 5 \\ \cline{1-2} \end{array}$$

with integer entries inside the boxes indicating that first row boxes are totally symmetric under exchange. Similarly, the boxes on the second row are also totally symmetric. The exchanges of boxes within the first column as well as exchange of boxes within the second column are each totally antisymmetric. Note that there is no symmetry between boxes numbered 3 and boxes numbered 5 belonging to a different row and a different column.

The dimension $d_Y$ of the irreducible representation $Y$ can be determined by counting the possible options of putting integer entries on the boxes such that they are increasing along their row or their column. Such a prescription incorporates the symmetric or antisymmetric or no-symmetry nature of the given Young diagram $Y$. The following example will illustrate the prescription giving dimension $d_Y$.

**Example 50.** Determine the dimensions of the irreducible representation $Y \in \mathfrak{S}(3)$ whose Young tableau is

We can place integers in the boxes following the above mentioned prescription

$$\begin{array}{|c|c|} \hline 1 & 2 \\ \hline 3 \\ \cline{1-1} \end{array} \,, \quad \begin{array}{|c|c|} \hline 1 & 3 \\ \hline 2 \\ \cline{1-1} \end{array} .$$

Thus there are only two possibilities indicating that the dimension $d_Y = 2$.

This procedure can be done for any $Y \in \mathfrak{S}(n)$ but the method may become tedious for large $n$. There is an alternative formula giving the dimensions $d_Y$:

$$d_Y = n! / \prod_{i \in Y} h_i, \tag{6.2.2}$$

where $h_i$ denotes the hook length of a box $i$ in the Young diagram $Y$. Basically, $h_i$ counts the number of boxes to the right of the box $i$ along the same row plus the number of boxes below the box $i$ along the column plus one (for the box $i$).

In the above example, the three hook lengths associated with the three boxes of $Y$ will be $h_1 = 3$, $h_2 = 1$, $h_3 = 1$ resulting in $d_Y = 3!/(3 \times 1 \times 1) = 2$ in agreement with Example 50.

The irreducible representations $Y$ of $\mathfrak{S}(n)$ has a close resemblance to the irreducible representation of $SU(n)$ but there are also differences. This naturally leads us to discuss $SU(N)$ group derived from the Young diagram approach.

## $SU(N)$ Young tableau

We will now briefly review the Young diagrams $Y$ depicting irreducible $SU(N)$ representations. Here $N$ denotes the number of possible states which we can place in every box of $Y$. Remember that the entry in each box of a Young diagram can be any of the $N$ values (with or without repetition) such that the properties of symmetric nature along the horizontal direction and antisymmetric nature along the vertical direction are maintained.

Suppose we try to place these states in a vertical column of $N$ boxes, we have only one possibility as the states along a column must be antisymmetric. Hence, one or more vertical columns of $N$ boxes denote a singlet or trivial representation. Unlike the $\sum_i \lambda_i = n$ constraint in the Young diagram $Y$ of the permutation group $\mathfrak{S}(n)$, there is no such restriction on the total number of boxes of $Y$ denoting non-trivial irreducible representation of $SU(N)$. However, the number of rows must not exceed $N-1$ for non-trivial irreducible representation. The dimension $d_Y$ for any $SU(N)$ Young diagram can be determined through combinatorial consideration of symmetric, antisymmetric and mixed symmetry properties of $N$ variables.

**Example 51.** Obtain the dimension of the irreducible representation $Y \in SU(3)$ whose Young diagram is ⊞.

The $SU(N=3)$ rule here is to put three states $u$, $d$, $s$ in the boxes such that repetition along the row is allowed but forbidden along the column. Hence the possible states are

| u | u |   | u | u |   | d | d |   | d | d |   | s | s |   | s | s |   | u | d |   | u | s |
|---|---|---|---|---|---|---|---|---|---|---|---|---|---|---|---|---|---|---|---|---|---|---|
| d |   |   | s |   |   | u |   |   | s |   |   | u |   |   | d |   |   | s |   |   | d |   |

giving dimension $d_Y = 8$.

Alternatively, there is a formula involving hook number

$$d_Y = Nr_Y / \prod_{i \in Y} h_i, \qquad (6.2.3)$$

where the numerator $Nr_Y$ is calculated as follows: place $N$, $N+1$, along the boxes in the first row and decreasing those integers in steps of one along the vertical columns. Then $Nr_Y$ is the product of all the integers in the boxes of $Y$.

The integer entries for the above example will be $\begin{array}{|c|c|} \hline 3 & 4 \\ \hline 2 \\ \cline{1-1} \end{array}$, giving $Nr_Y = 3 \times 4 \times 2 = 24$ and its hook length as discussed in Example 50 is 3. Hence, we get dimension $d_Y = 24/3 = 8$ which is in agreement with that obtained in Example 51.

**Example 52.** Deduce the Young diagram $Y \in SU(2)$ corresponding to the irreducible representation $V^j$ whose dimension is $2j+1$.

Recall that the non-trivial irreducible representation of $SU(2)$ can be shown by a single row Young diagram $\begin{array}{|c|c|c|c|c|c|} \hline a_1 & a_2 & \cdot & \cdot & \cdot & a_m \\ \hline \end{array}$ where $a_i$'s can be either $u$

or $d$. Note that diagrams with one or more columns of two boxes $\begin{array}{|c|}\hline a_1 \\ \hline b_1 \\ \hline\end{array}$ are trivial one-dimensional representations because if $a_1$ is chosen as $u$ then $b_1$ is necessarily $d$. Hence, for an irreducible representation $Y = \begin{array}{|c|c|c|c|c|c|}\hline a_1 & a_2 & . & . & . & a_m \\ \hline\end{array}$ of $SU(2)$, the dimension can be calculated using formula (6.2.3). Placing integers $N = 2$ in the first box and increasing them one by one along the row, we can determine the dimension to be

$$d_Y = Nr_Y / \prod_{i \in Y} h_i = (2 \times 3 \times \ldots m+1)/(m)! = m+1.$$

This implies that the irreducible representation $V^j$, corresponding to angular momentum $j$, whose dimension is $2j+1$ can be presented as the following Young tableau: $\begin{array}{|c|c|c|c|c|c|}\hline a_1 & a_2 & . & . & . & a_{2j} \\ \hline\end{array}$. □

This has been a quick overview of Young tableau presentation of irreducible representation of $SU(N)$. We can now understand the diagrammatic construction of tensor product $Y_1 \otimes Y_2$ of two irreducible representations and their decomposition $\oplus Y$.

**Highest weight associated with $SU(N)$ Young diagrams**
In the previous chapter, we deduced the two $SU(3)$ fundamental weight vectors from the two simple root vectors. The fundamental weight vector $\boldsymbol{\mu}^{(1)}$ is the highest weight state of representation $\square$ and the fundamental weight $\boldsymbol{\mu}^{(2)}$ vector is the highest weight state of representation $\begin{array}{|c|}\hline \\ \hline \\ \hline\end{array}$. Any arbitrary representation of $SU(3)$ will have $a_1$ single boxes and $a_2$ double vertical boxes. For example, the Young diagram has $a_1 = 5$ and $a_2 = 2$. The highest weight vector of such a $SU(3)$ representation will be $\Lambda = a_1 \boldsymbol{\mu}^{(1)} + a_2 \boldsymbol{\mu}^{(2)}$. The Young diagram can be equivalently denoted as $(a_1, a_2)$.

This procedure can be generalized to $SU(N)$ group as follows: there will be $N-1$ component fundamental weight vectors $\boldsymbol{\mu}^{(i)}$'s where $i = 1, 2, \ldots N-1$ whose corresponding Young diagrams will be $\square, \begin{array}{|c|}\hline\\\hline\end{array}, \begin{array}{|c|}\hline\\\hline\\\hline\end{array}, \ldots$ respectively. Further arbitrary $SU(N)$ representation can be written as $(a_1, a_2, \ldots, a_{N-1})$ whose equivalent Young diagram will have $a_1$ single boxes, $a_2$ double vertical boxes and so on, and the corresponding highest weight will be

$$\Lambda = a_1 \boldsymbol{\mu}^{(1)} + a_2 \boldsymbol{\mu}^{(2)} \ldots + a_{N-1} \boldsymbol{\mu}^{(N-1)}.$$

# Further Applications

**Diagrammatic understanding of tensor product**

Let us explain this aspect through the following simple example:

$$\underbrace{\boxed{a_1\,a_2}}_{Y_1} \otimes \underbrace{\boxed{b_1\,b_2\,b_3}}_{Y_2} = \boxed{a_1\,a_2\,b_1\,b_2\,b_3} \oplus \boxed{\begin{array}{cccc}a_1 & a_2 & b_1 & b_2 \\ b_3 & & & \end{array}} \oplus$$

$$\boxed{\begin{array}{ccc}a_1 & a_2 & b_1 \\ b_2 & b_3 & \end{array}}$$

(6.2.4)

where we have placed $a_i$'s in the boxes of $Y_1$ and $b_i$'s in the boxes of $Y_2$. This will keep track of decomposition possibilities $\oplus_\alpha Y_\alpha$ as depicted above on the right hand side. Note that the symmetric nature amongst boxes with entries $a_i$'s and amongst boxes with entries $b_i$'s are maintained in the irreducible representations $Y$. Also observe that the boxes with entries $a_i$ can be symmetric or antisymmetric with respect to boxes with entries $b_i$ in the irreducible representations $Y$.

**Example 53.** Work out the dimensions for the above Young diagrams $Y_1, Y_2, Y \in SU(4)$ and verify

$$d_{Y_1} d_{Y_2} = \sum_{Y \in Y_1 \times Y_2} d_Y.$$

The dimensions of representations of $SU(4)$ (6.2.3):

$$d_{\boxed{4\,5}} = (4 \times 5)/(1 \times 2) = 10,$$

$$d_{\boxed{4\,5\,6}} = (4 \times 5 \times 6)/(1 \times 2 \times 3) = 20.$$

Hence LHS: $d_{Y_1} d_{Y_2} = 200$. In the similar fashion, we can work out

$$d_{\boxed{4\,5\,6\,7\,8}} = 56,$$

$$d_{\boxed{\begin{array}{cccc}4&5&6&7\\3&&&\end{array}}} = (3 \times 4 \ldots 7)/(1 \times 5 \times 3 \times 2 \times 1) = 84 \text{ and}$$

$$d_{\boxed{\begin{array}{ccc}4&5&6\\3&4&\end{array}}} = 1440/(2 \times 4 \times 3 \times 1) = 60.$$

Thus RHS: $\sum_{Y \in Y_1 \times Y_2} d_Y = 56 + 84 + 60 = 200.$ □

The tensor product (6.2.4) and $SU(N)$ decomposition can be further simplified for the $SU(2)$ group as vertical column(s) with two boxes are trivial:

$$\boxed{a_1\ a_2} \otimes \boxed{b_1\ b_2\ b_3} = \boxed{a_1\ a_2\ b_1\ b_2\ b_3} \oplus \boxed{a_2\ b_1\ b_2} \oplus \boxed{b_1} \tag{6.2.5}$$

which agrees with the tensor product Equation 6.2.1 $V^1 \otimes V^{3/2} = \oplus_{j=1/2}^{5/2} V^j$ where $V^j$ denotes single row Young tableau with $2j$ boxes. Just like the basis states $|j,m\rangle$ belong to $V^j$, we could also have rank $k$ irreducible tensors $O(k,q)$ whose transformation properties are similar to the state $|k,q\rangle$. For example, vector $\vec{A}$ belongs to rank 1 irreducible tensor. Tensor product of two vectors $\vec{A}$ and $\vec{B}$ whose decomposition will be exactly like Equation 6.2.1:

$$\vec{A} \otimes \vec{B} = \oplus_{k=0}^{2} O(k,q).$$

Comparing the form of spherical harmonics $Y_m^{\ell=1}(\theta, \phi)$ with the components of position vector, the vector $\vec{A}$ can be rewritten in the spherical rank 1 tensor form as

$$A(1,0) = A_z, \quad A(1,\pm 1) = \mp \frac{A_x \pm i A_y}{\sqrt{2}}. \tag{6.2.6}$$

We had elaborated the projector method of obtaining binary and tertiary basis from tensor product in the discrete group context. In similar fashion, the states $|j,m\rangle$ belonging to the irreducible space $V^j \in V^{j_1} \otimes V^{j_2}$ must be rewritable as the linear combination of tensor product states $|j_1,m_1\rangle|j_2,m_2\rangle$. The projection method for tensor product of $SU(2)$ representations is the focus of the following section.

### 6.2.2 $SU(2)$ Clebsch–Gordan matrix

The decomposition into irreducible component states $|j,m\rangle \in V^j$ from the tensor product of states $|j_1,m_1\rangle \otimes |j_2,m_2\rangle$ involves a matrix $C^{j,m}_{j_1,m_1;j_2,m_2}$ called Clebsch–Gordan (CG) coefficient matrix:

$$|j,m\rangle = \sum_{m_1 m_2} C^{j,m}_{j_1,m_1;j_2,m_2} |j_1,m_1\rangle \otimes |j_2,m_2\rangle, \tag{6.2.7}$$

where CG coefficients are non-zero if and only if $|j_1-j_2| \leq j \leq j_1+j_2$ and $m = m_1 + m_2$. These CG coefficients are available in the form of tables. We can also obtain these coefficients by assuming CG coefficient of the maximum angular momentum state, $|j_{\max} = j_1+j_2, m = j_1+j_2\rangle$ (also called as highest weight state), as $C^{j_1+j_2,j_1+j_2}_{j_1,j_1;j_2,j_2} = 1$, i.e.,

$$|j_1+j_2, j_1+j_2\rangle = |j_1,j_1\rangle|j_2,j_2\rangle. \tag{6.2.8}$$

## Further Applications

Recall the action of the $\mathfrak{su}(2)$ generators $J_\pm$, $J_z$ on the states discussed in the previous chapter:

$$J_\pm |j, m\rangle = \hbar \sqrt{(j \mp m)(j \pm m + 1)} |j, m \pm 1\rangle; \quad J_z |j, m\rangle = \hbar m |j, m\rangle. \tag{6.2.9}$$

Incidentally, $\mathfrak{su}(2)$ generators on the tensor product state $|j_1, m_1\rangle |j_2, m_2\rangle$ can be deduced from the tensor product of the infinitesimal group elements : $(\mathbb{I} + \sum_{a=1}^{3} J_a^{(1)} \delta\theta^a) \otimes (\mathbb{I} + \sum_{a=1}^{3} J_a^{(2)} \delta\theta^a)$ where the superscript on the generators keeps track of the angular momentum. Keeping up to order $\mathcal{O}(\delta\theta^2)$, the tensor product simplifies to

$$\mathbb{I} + \left[ \left( \sum_{a=1}^{3} J_a^{(1)} \theta^a \right) \otimes \mathbb{I} + \mathbb{I} \otimes \left( \sum_{a=1}^{3} J_a^{(2)} \theta^a \right) \right] + \mathcal{O}(\delta\theta^2),$$

indicating that the generators acting on tensor product state must be

$$[J \otimes \mathbb{I} + \mathbb{I} \otimes J] |j_1, m_1\rangle \otimes |j_2, m_2\rangle.$$

Applying these $\mathfrak{su}(2)$ generators on both the LHS and RHS of Equation 6.2.8 leads to the following:

$$J_- |j_1 + j_2, j_1 + j_2\rangle = \sqrt{2j_1 + 2j_2} |j_1 + j_2, j_1 + j_2 - 1\rangle. \tag{6.2.10}$$

The action of these generators on the tensor product state of RHS of Equation 6.2.8 will be

$$\{J_- |j_1, j_1\rangle\} |j_2, j_2\rangle + |j_1, j_1\rangle \{J_- |j_2, j_2\rangle\} = \sqrt{2j_1} |j_1, j_1 - 1\rangle |j_2, j_2\rangle + \sqrt{2j_2} |j_1, j_1\rangle |j_2, j_2 - 1\rangle. \tag{6.2.11}$$

From the two equations given above (Equations 6.2.10 and 6.2.11), we obtain

$$|j_1 + j_2, j_1 + j_2 - 1\rangle = \sqrt{\frac{j_1}{j_1 + j_2}} |j_1, j_1 - 1\rangle |j_2, j_2\rangle + \sqrt{\frac{j_2}{j_1 + j_2}} |j_1, j_1\rangle |j_2, j_2 - 1\rangle. \tag{6.2.12}$$

Comparing the above equation with Equation 6.2.7, we can read off CG coefficients

$$C^{j_1+j_2,j_1+j_2-1}_{j_1,j_1;j_2,j_2-1} = \sqrt{\frac{j_2}{j_1 + j_2}}, \quad C^{j_1+j_2,j_1+j_2-1}_{j_1,j_1-1;j_2,j_2} = \sqrt{\frac{j_1}{j_1 + j_2}}. \tag{6.2.13}$$

This procedure can be similarly continued acting $J_-$ on Equation 6.2.12.

**Example 54.** Work out the CG coefficients $C^{1/2,1/2}_{1/2,1/2;1,0}$, $C^{1/2,1/2}_{1/2,-1/2;1,1}$. From Equation 6.2.12, the state $|3/2, 1/2\rangle$ is

$$|3/2, 1/2\rangle = \sqrt{\frac{1}{3}}|1/2, -1/2\rangle|1, 1\rangle + \sqrt{\frac{2}{3}}|1/2, 1/2\rangle|1, 0\rangle.$$

We need to write the state $|1/2, 1/2\rangle$ to obtain the CG coefficients $C^{1/2,1/2}_{1/2,1/2;1,0}$, $C^{1/2,1/2}_{1/2,-1/2;1,1}$. This state must be orthogonal to $|3/2, 1/2\rangle$. Hence

$$|1/2, 1/2\rangle = \sqrt{\frac{1}{3}}|1/2, -1/2\rangle|1, 1\rangle - \sqrt{\frac{2}{3}}|1/2, 1/2\rangle|1, 0\rangle,$$

which leads to the CG coefficients:

$$C^{1/2,1/2}_{1/2,1/2;1,0} = -\sqrt{\frac{2}{3}}, \quad C^{1/2,1/2}_{1/2,-1/2;1,1} = \sqrt{\frac{1}{3}}. \qquad \square$$

The CG method, which we described for the tensor product of any two states, is also applicable to the tensor product of two irreducible rank tensors $O(k_1, q_1)$, $O(k_2, q_2)$ or the tensor product of $O(k, q)$ with state $|j_1, m_1\rangle$. Similar to the action of $\mathfrak{su}(2)$ generators on states $|j, m\rangle$ in Equation 6.2.9, their action on the irreducible tensors can be shown to satisfy:

$$[J_z, O(k, q)] = \hbar q\, O(k, q), \qquad (6.2.14)$$

$$[J_\pm, O(k, q)] = \hbar\sqrt{k(k+1) - q(q\pm 1)}\, O(k, q\pm 1). \qquad (6.2.15)$$

**Example 55.** Using the familiar transformation of rank one tensor $\vec{A}$ under infinitesimal rotation $R$ by angle $\delta\theta$ about axis $\hat{n}$ to the quantum mechanical operator transformation

$$\vec{A}' = R\vec{A} = \vec{A} + \delta\theta \hat{n} \times \vec{A} = U_R \vec{A} U_R^\dagger,$$

prove Equations 6.2.14 and 6.2.15 where $U_R = \left(1 + \frac{i\delta\theta}{\hbar}\hat{n}\cdot\vec{J}\right)$.

Equating $\mathcal{O}(\delta\theta)$ terms in the above equation, we get

$$\frac{i}{\hbar}[\hat{n}\cdot\vec{J}, \vec{A}] = \hat{n} \times \vec{A},$$

which simplifies for components of $\vec{A}$. For instance, the above equation for $A_x$ turns out to be

$$[A_x, J_x] = 0\,;\ [A_x, J_y] = i\hbar A_x\,;\ [A_x, J_z] = -i\hbar A_y,$$

which implies the following compact form

$$[J_i, A_j] = \epsilon_{ijk} i\hbar A_k,$$

leading to obtaining the algebra for $A(1, q)$ (6.2.6):

$$[J_z, A(1, q)] = \hbar q A(1, q) \,;\, [J_\pm, A(1, q)] = \hbar \sqrt{2 - q(q \pm 1)} A(1, q \pm 1). \qquad \square$$

**Example 56.** Construct the rank 2 tensor component $T(2, 1)$ using two rank one tensors $A(1, q_1)$ and $B(1, q_2)$.

Following the Clebsch–Gordan procedure for states, we start with the maximum $q$ tensor as

$$T(2, 2) = A(1, 1)B(1, 1),$$

and applying $J_-$ on both LHS as well as RHS will give

$$T(2, 1) = \frac{1}{\sqrt{2}} [A(1, 1)B(1, 0) + A(1, 0)B(1, 1)]. \qquad \square$$

Even though we confined ourselves to the construction of $SU(2)$ CG matrix and angular momentum states, the method can be extended to $SU(N)$ starting with the highest weight vector state $|\Lambda, \Lambda\rangle$ and applying $su(N)$ generators $E_{\pm\alpha}$, $H_i$ associated with simple root vector $\alpha$.

Recall our discussion on highest weight vector states of the irreducible representation of $SU(N)$. For the Young diagram depicted: $\in SU(3)$, the highest weight vector will be

$$\Lambda = 3\mu^{(1)} + 2\mu^{(2)},$$

where $\mu^{(1)}$, $\mu^{(2)}$ are the two fundamental weights of $SU(3)$ discussed in the previous chapter (see Example 48).

Similar to the $SU(2)$ CG construction, we can take the tensor product of two $SU(3)$ irreducible representations of highest weights $|\Lambda_1, \Lambda_1\rangle \otimes |\Lambda_2, \Lambda_2\rangle$ to give highest weight vector $\Lambda = \Lambda_1 + \Lambda_2$. The other weight vector states are obtained by applying $E_{-\alpha^{(1)}}$, $E_{-\alpha^{(2)}}$, $E_{-\alpha^{(3)}}$ on both sides as follows:

$$E_{-\alpha^{(i)}} |\Lambda, \Lambda\rangle \propto |\Lambda, \Lambda - \alpha^{(i)}\rangle = [E_{-\alpha^{(i)}} |\Lambda_1, \Lambda_1\rangle] |\Lambda_2, \Lambda_2\rangle + |\Lambda_1, \Lambda_1\rangle [E_{-\alpha^{(i)}} |\Lambda_2, \Lambda_2\rangle].$$

The proportionality factor can be deduced by the action of the corresponding $\mathfrak{su}(2)$ subalgebra generators $J_3^{(\alpha)}$, $J_-^{(\alpha)}$ on the highest weight states:

$$J_3^{(\alpha)}|\Lambda, \Lambda\rangle = \frac{(\Lambda.\alpha)}{|\alpha|^2}|\Lambda, \Lambda\rangle,$$

$$J_-^{(\alpha)}|\Lambda, \Lambda\rangle = \frac{\sqrt{[(\Lambda + \Lambda).\alpha][(\Lambda - \Lambda).\alpha + 1]}}{|\alpha|}|\Lambda, \Lambda - \alpha\rangle.$$

For the lower weight states $|\Lambda, \Lambda - m(\alpha)^{(1)} - n\alpha^{(2)}\rangle$ where $m$, $n$ are integers, the $J_3^{(\alpha)}$, $J_\pm^{(\alpha)}$ operator action will be

$$J_3^{(\alpha)}|\Lambda, \Lambda - m\alpha^{(1)} - n\alpha^{(2)}\rangle = \frac{\left(\Lambda - m\alpha^{(1)} - n\alpha^{(2)}\right).\alpha}{|\alpha|^2}|\Lambda, \Lambda - m\alpha^{(1)} - n\alpha^{(2)}\rangle, \quad (6.2.16)$$

$$J_\mp^{(\alpha)}|\Lambda, \Lambda - m\alpha^{(1)} - n\alpha^{(2)}\rangle = \frac{1}{|\alpha|}\sqrt{[\{\Lambda \pm (\Lambda - m\alpha^{(1)} - n\alpha^{(2)})\}.\alpha]}$$

$$\sqrt{[\{\Lambda \mp (\Lambda - m\alpha^{(1)} - n\alpha^{(2)})\}.\alpha + 1]}|\Lambda, \Lambda - m\alpha^{(1)} - n\alpha^{(2)} \mp \alpha\rangle.$$

This CG construction of tensor product of weight vector states will be useful for determining the states of hadrons which are bound states of fundamental quarks. These hadronic states in the continuous group context are similar to the binary or tertiary basis discussed in the discrete group context. We will illustrate in Section 6.4.1 for a bound state when we discuss hadrons in particle physics.

In the following subsection, we will look at the selection rules for transition from the initial state to the final state due to interactions respecting $SU(2)$ symmetry. In Chapter 4, we had deduced that the matrix elements in the discrete symmetry situation would be non-zero if the tensor product of the irreducible representations corresponding to basis states and operators turned out to be trivial representation. In the same fashion, the matrix elements $\langle \beta j_f, m_f|O(k, q)|\alpha j_i, m_i\rangle$ between initial state $\alpha$ with angular momentum $j_i$ and final state $\beta$ angular momentum $j_f$ due to interactions, like electric dipole moment interaction or quadrapole moment interaction and suchlike, denoted by the irreducible rank $k$ tensor operator $O(k, q)$ can be deduced to be vanishing or non-vanishing. This is the content of Wigner–Eckart theorem which we will present with proof.

### 6.2.3 Wigner–Eckart theorem

The theorem states that the matrix elements of the irreducible tensor operator between two angular momentum states consists of product of two terms

$$\langle \beta j_f, m_f|O(k, q)|\alpha j_i, m_i\rangle = \langle \beta j_f||O(k)||\alpha j_i\rangle C^{j_f, m_f}_{j_i, m_i; k, q},$$

where the first factor is called reduced matrix element which depends on the quantum states $|\alpha j_i\rangle$, $|\beta j_f\rangle$ and the rank of the tensor operator whereas the second factor is purely geometrical given by the CG coefficient. Even though we need experimental data to determine the reduced matrix element, we can at least rule out (using CG coefficients) whether the process is allowed or forbidden in any system with rotational spherical symmetry: $SU(2)$.

From Equation 6.2.14 we know that

$$\langle \beta j_f, m_f | [j_z, O(k,q)] - \hbar q O(k,q) | \alpha j_i, m_i \rangle = 0,$$

leading to the constraint $m_f - m_i = q$ for non-zero $\langle \beta j_f, m_f | O(k,q) | \alpha j_i, m_i \rangle$ when we expand the commutator. This is one of the properties of the CG coefficient. Similarly, we can impose the following constraint using Equation 6.2.15:

$$\langle \beta j_f, m_f | [j_\pm, O(k,q)] - \hbar \sqrt{(k \mp q) - (k \pm q + 1)} O(k, q \pm 1) | \alpha j_i, m_i \rangle = 0.$$

Again expanding the commutator, the equation resembles the recursion relation obeyed by the Clebsch–Gordan coefficients. Thus we conclude that

$$\langle \beta j_f, m_f | O(k,q) | \alpha j_i, m_i \rangle \propto C_{j_i, m_i; k, q}^{j_f, m_f},$$

and the proportionality constant will be independent of $m_f, m_i, q$ proving the Wigner–Eckart Theorem. □

**Example 57.** Given that the diagonal matrix element of $\langle \alpha j, m' = j | Q(2,0) | \alpha j, m = j \rangle = K$, determine the matrix element of

$$\langle \alpha j, m' | Q(2,-2) | \alpha j, m = j \rangle.$$

Using Wigner–Eckart theorem, $\langle \alpha j, m' = j | Q(2,0) | \alpha j, m = j \rangle = K = \langle \alpha j || Q(2) || \alpha j \rangle C_{j,j;2,0}^{j,j}$. This determines the reduced matrix element in terms of $K$ and Clebsch–Gordan coefficient:

$$\langle \alpha j || Q(2) || \alpha j \rangle = K / C_{j,j;2,0}^{j,j}.$$

Once we know the reduced matrix element then

$$\langle \alpha j, m' | Q(2,-2) || \alpha j, m = j \rangle = \left( K C_{j,j;2,-2}^{j, m' = j-2} \right) / C_{j,j;2,0}^{j,j}.$$

Thus the theorem is useful in determining the matrix elements of other components of $Q(2, q)$ from the experimental result for one component. □

## 6.3 Elementary Particles in Nuclear and Particle Physics

In this section, we will focus on the applications of $SU(2)$ and $SU(3)$ symmetries in nuclear and particle physics. The periodic table of elementary particles is broadly classified as *hadrons and leptons*. The electromagnetic and nuclear interactions amongst these particles are mediated by *new particles* which are referred to as *force carriers*. For instance, *photon* $\gamma$ is the mediator of electromagnetic interaction. Similarly, there are eight gluons $g_i$'s which are mediators of strong nuclear interaction and three vector bosons $W\pm$, $Z$ mediating weak nuclear interaction. In order to give masses to these particles, the Higgs particle was theoretically proposed and recently discovered in the Large Hadron Collider experiments at CERN, Geneva.

Hadrons are further subclassified as *baryons* and *mesons*. The protons and neutrons inside the atomic nucleus are examples of baryons. Surprisingly, the nuclear magnetic moment of the proton and neutron are proportional to nuclear magneton but the proportionality constant is a real number and not an integer. Hence the protons and neutrons cannot be elementary but must have a structure. The other massive baryons like $\triangle$, $\Sigma$, ... appear in the collision experiments involving high energy proton beams. Similar to the Rutherford scattering experiment which suggested a massive nucleus inside the atom, deep inelastic scattering experiments of baryons indicated that they are bound states of three quarks. Unlike electrons, protons and neutrons, free quarks are not seen in nature.

In the language of group theory, we will say that baryons can be obtained by taking the tensor product of three fundamental representations associated with the quarks. Unlike baryons, mesons are the bound states of quark and antiquark pairs. For example, pions $\pi$ and $\rho$ particle are mesons.

Murray Gell-Mann was influenced by the pattern diagrams emerging from plotting baryons and mesons based on their mass $m$ and charge $Q$ for a fixed angular momentum $J$ (some patterns resemble an eight-fold path). Importantly, he proposed the *quark model* to reproduce the observed patterns. The readers can actually visualize that these pattern diagrams are nothing but the weight diagrams discussed in Chapter 5 for $SU(3)$ Lie group.

From the experimentally observed strong nuclear scattering or decay process of hadrons, theoreticians postulated that there must be an additional conservation law (isospin conservation) besides linear momentum, energy and angular momentum conservation laws. Similar to Wigner–Eckart theorem, discussed in the context of $\mathfrak{su}(2)$ angular momentum states and operators, there will be selection rules for such scattering or decay processes obeying isospin conservation.

We will briefly discuss selection rules of strong scattering and decay processes involving hadrons in the following subsection and then elaborate the Gell-Mann quark model, in Section 6.4, to understand the theoretical construction of observed baryons and mesons and the pattern diagrams.

## Further Applications

### 6.3.1 Isospin symmetry

There is a quantity called *isospin* $\vec{I}$ in nuclear and particle physics whose Lie algebra

$$[I_i, I_j] = i\epsilon_{ijk} I_k$$

is exactly similar to angular momentum algebra. Interestingly any decay or scattering process involving hadrons $A \to B + C$ must conserve isospin. In other words, isospin symmetry is respected in such strong interaction processes. The tensor operators describing such scattering/decay processes are scalar operators $S(0, 0)$. We can now study allowed and forbidden processes using Wigner–Eckart theorem.

The isospin I and the z-component of isospin $I_z$ for few baryons and mesons which are needed to determine Clebsch–Gordan coefficients are as follows:

delta particles $\Delta$: $I = 3/2, I_z = +3/2, 1/2, -1/2, -3/2$ for $\Delta^{++}, \Delta^+, \Delta^0, \Delta^-$,

nucleons N: $I = 1/2, I_z = 1/2, I_z = -1/2$ for $p$ and $n$,

pi meson $\pi$: $I = 1, I_z = +1, 0, -1$ for $\pi^+, \pi^0$ and $\pi^-$,

rho meson $\rho$: $I = 1, I_z = +1, 0, -1$ for $\rho^+, \rho^0$ and $\rho^-$.

With this data, let us attempt to figure out the allowed and forbidden processes of strongly interacting baryons and mesons.

**Example 58.** Show that the the ratio of the decay rate between the two strong decay processes respecting isospin symmetry is

$$\frac{\Gamma[\Delta^+ \to p\pi^0]}{\Gamma[\Delta^+ \to n\pi^+]} = \frac{|\langle p\pi^0|S(0,0)|\Delta^+\rangle|^2}{|\langle n\pi^+|S(0,0)|\Delta^+\rangle|^2} = 2.$$

Using the Wigner–Eckart theorem, the decay process $\Delta^+ \to p\pi^0$ is given by the matrix element

$$\langle p\pi^0|S(0,0)|\Delta^+\rangle \propto \langle p\pi^0|3/2,1/2\rangle = C_{1/2,1/2;1,0}^{3/2,1/2}, \qquad (6.3.1)$$

where $|p\pi^0\rangle \equiv |1/2,1/2;1,0\rangle = |1/2,1/2\rangle|1,0\rangle$. Similarly the decay amplitude of $\Delta^+ \to n\pi^+$ is given by

$$\langle n\pi^+|S(0,0)|\Delta^+\rangle \propto \langle n\pi^+|3/2,1/2\rangle = C_{1/2,-1/2;1,1}^{3/2,1/2}. \qquad (6.3.2)$$

The Clebsch–Gordan decomposition of the state $|3/2,1/2\rangle$ obtained from the tensor product of $I_1 = 1/2 \otimes I_2 = 1$ will be

$$|3/2,1/2\rangle = \sqrt{\frac{2}{3}} \underbrace{|1/2,1/2;1,0\rangle}_{|p,\pi^0\rangle} + \sqrt{\frac{1}{3}} \underbrace{|1/2,-1/2;1,1\rangle}_{|n\pi^+\rangle}.$$

Using the above CG coefficients, the ratio of Equations 6.3.1 and 6.3.2 gives

$$\frac{\Gamma[\triangle^+ \to p\pi^0]}{\Gamma[\triangle^+ \to n\pi^+]} = \frac{\left(\sqrt{\frac{2}{3}}\right)^2}{\left(\sqrt{\frac{1}{3}}\right)^2} = 2.$$

$\square$

## 6.4 Quark Model

The angular momentum of the baryons are either $J = 3/2$ or $J = 1/2$. Further, the observed magnetic moments of protons are not integral multiples of nuclear magneton $\mu_N = e\hbar/(2m_p)$ indicating the baryons are not elementary particles but composites made up of $n$ number of quarks which are spin $1/2$ particles.

The angular momentum state of these spin $1/2$ quarks belong to the two-dimensional space $V^{1/2}$. Baryonic states belong to the irreducible spaces obtained from the tensor product of these $n$ quarks:

$$\underbrace{V^{1/2} \otimes V^{1/2} \otimes \ldots V^{1/2}}_{n} = V^{n/2} \oplus V^{n/2-1} \oplus \ldots$$

The angular momentum of the irreducible representations, denoting baryonic states, must have either spin $3/2$ or spin $1/2$. Equating highest spin $n/2$ of the irreducible space $V^{n/2}$ to spin $3/2$, we deduce that the baryons must be made up of $n = 3$ quarks.

Even though we have accounted for the observed angular momentum of baryons, we still need a radical idea to explain the reason for eight baryons including nucleons to have $J = 1/2$ and the rest of the baryons including $\triangle$ particles to have $J = 3/2$. Figure 6.4.1(b) shows the plots of $J = 1/2$ baryons resembling eight-fold path diagram and Figure 6.4.1(a) for $J = 3/2$ baryons. Note that the dotted lines indicate the line of same charge $Q$ and the horizontal lines have same isospin I and decreasing $I_z$ components in steps of one from right to left. Interestingly, Gell-Mann's mathematical formulations used Lie groups to produce the observed plots. This is famously known as Gell-Mann's quark model for which he received the Nobel Prize in the year 1969. We will explain the rescaling of the vertical axis and the definition of $Y$ in the next subsection.

Further Applications    133

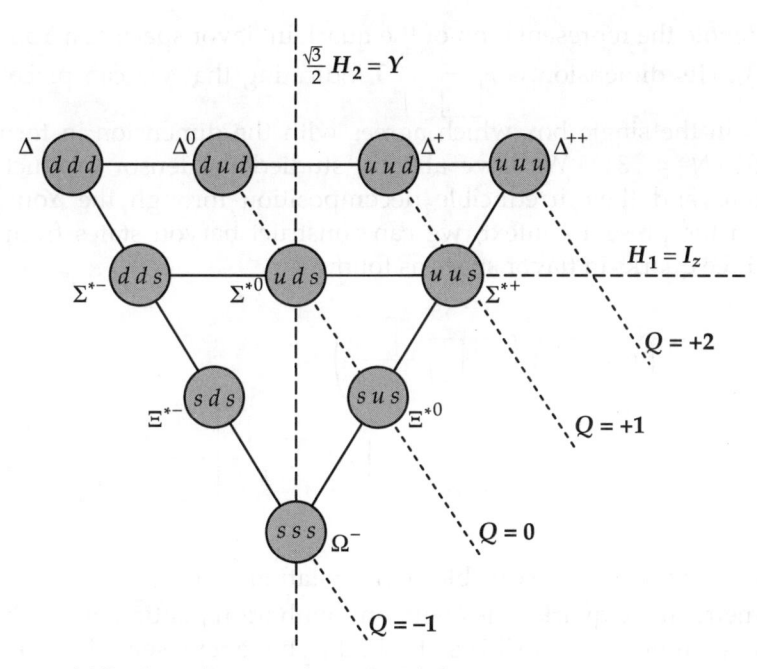

(a) Decimet diagram of spin $J = 3/2$ baryons

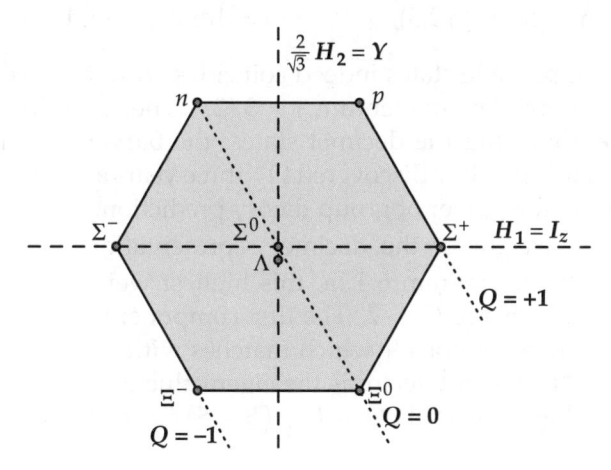

(b) Octet diagram of spin $J = 1/2$ baryons

**Figure 6.4.1**    Quark Model

## 6.4.1 $SU(3)$ group approach quark model

Gell-Mann assumed that the quarks come in three flavors $u$, $d$, $s$ (up, down, strange quarks). In the flavor space, $(u, d)$ are states having isospin $\mathbf{I} = 1/2$ and have the z-component of isospin as $+1/2, -1/2$ respectively, whereas $s$ has isospin $\mathbf{I} = 0$.

We can denote the representation of the quark in flavor space as a Young diagram $\square \in SU(3)$. Its dimension is $d_{\square} = 3$ indicating that we can place any of the three-flavors in the single box which agrees with the dimension in formula (6.2.3) for group $SU(N = 3)$. We have already studied the tensor product of $SU(N)$ representations and their irreducible decomposition through the Young diagram approach. In the present context, we can construct baryon states from the tensor product of three quarks in flavor space as follows:

$$\square \otimes (\square \otimes \square) = \square \otimes \left(\square\square \oplus \begin{smallmatrix}\square\\\square\end{smallmatrix}\right) = \square\square\square \oplus \begin{smallmatrix}\square\square\\\square\end{smallmatrix} \oplus \begin{smallmatrix}\square\square\\\square\end{smallmatrix} \oplus \begin{smallmatrix}\square\\\square\\\square\end{smallmatrix} \tag{6.4.1}$$

The baryons belonging to irreducible representation $\square\square\square$ refers to all possible totally symmetric three quark states with and without repetition of the three flavors. Hence, the total number of possible states in the above representation will be 10 with the following possible entries in the boxes : *uuu*; *ddd*; *sss*; *uud*; *uus*; *ddu*; *dds*; *ssu*; *ssd*; *uds*. Using the dimension formula (6.2.3), $d_{\square\square\square} = 10$. We would like to emphasize that the pattern of the ten possible states indeed coincides with the observed pattern diagram of baryons with angular momentum $\mathbf{J} = 3/2$. When Gell-Mann proposed the $SU(3)$ quark model deducing the decimet states, the baryon $\Omega^-$ had not been discovered by experimentalists. They discovered $\Omega^-$ three years after the quark model hypothesis, demonstrating the power of group theory prediction.

Note that the highest weight of the decimet representation is $\Lambda = 3\mu^{(1)} = (3/2, 3\sqrt{3}/6)$. In the decimet diagram 6.4.1a, this highest weight state is the $\triangle^{++}$ baryon whose $I_z = 3/2$ and charge $Q = 2$. The first component of the weight vector is the eigenvalue of Cartan generator $H_1$ which matches with the $I_z = 3/2$ whereas the second component of the weight vector is the eigenvalue of Cartan generator $H_2$. Using the Gell-Mann–Nishijima formula $Q = I_z + (B + S)/2$, the following relation to $H_2$ can be deduced:

$$\frac{2}{\sqrt{3}} H_2 = 2(Q - I_z) \equiv Y = (B + S),$$

where $Y$ is referred to as *hypercharge* whose value can be deduced using the charge $Q$ and $z$-component of isospin of the particles. Equivalently, all baryons are assigned baryon number $B = +1$. If the baryons have strange quarks as constituents, these baryons have a non-zero strangeness $S$ quantum number. The $\triangle$ particles have $B = 1$, $S = 0$ whereas $\Sigma^{*+}$ has $B = 1$, $S = -1$. That is, for every $s$ quark, we associate $S = -1$. In terms of $B$, $S$, the hypercharge $Y = B + S$ can be obtained. Using the lowering operators $E_{-\alpha^{(i)}}$ on the highest weight vector state $|\Lambda, \Lambda = 3\mu^{(1)} \equiv \Delta^{++}\rangle$, we can obtain the remaining nine baryons in the deciment diagram. Hence, the deciment diagram is indeed a weight diagram with the $H_1 \equiv I_z$ and $H_2 = \sqrt{3}Y/2$ for the irreducible representation ☐☐ $\in SU(3)$. Following the relation of $H_1$ and $H_2$ to isospin and hypercharge, the $SU(3)$ defining representation weight diagram states discussed in Chapter 5 can be equivalently interpreted as $|\mu_1 \equiv u\rangle$, $|\mu_2 \equiv d\rangle$, $|\mu_3 \equiv s\rangle$.

We leave the readers to compute the dimensions of the other irreducible representation of $SU(3)$ in the tensor product ☐ ⊗ ☐ ⊗ ☐ and verify that the dimension of Equation 6.4.1 is

$$3 \otimes 3 \otimes 3 = 10 \oplus 8 \oplus 8 \oplus 1.$$

Note that there are two eight-dimensional representations from the group theory tensor product whereas there is only one octet weight diagram for baryons. The weight vector states of each irreducible representation can be obtained using the CG method and shown that they are orthonormal to each other. In fact, the experimental value of the magnetic moment data can be reproduced only for a suitable linear combination of the two eight-dimensional representation. This experimentally stringent requirement justifies one octet diagram of baryons. Interested readers can refer to any elementary particle physics textbook for details.

Let us now work out a simple example to determine $SU(3)$ CG coefficients.

**Example 59.** For a di-quark bound state, determine the six weight vector states associated with the Young diagram ☐☐ $\in SU(3)$ using the CG construction.

The highest weight vector state is $|2\mu^{(1)}, 2\mu^{(1)}\rangle \equiv |\boxed{u\,u}\rangle = |uu\rangle$. We can now apply the lowering operators $J_-^{(\alpha^{(1)})}$ on the highest weight state. Before we do that, we know that

$$(2\alpha^{(1)}.2\mu^{(1)})/|\alpha^{(1)}|^2 = q = 2.$$

Thus there will be two weight vector states

$$|2\mu^{(1)}, 2\mu^{(1)} - \alpha^{(1)}\rangle, \ |2\mu^{(1)}, 2\mu^{(1)} - 2\alpha^{(1)}\rangle.$$

Applying the lowering operator (6.2.16):

$$J_-^{(\alpha^{(1)})}|2\mu^{(1)}, 2\mu^{(1)}\rangle = \sqrt{2}|2\mu^{(1)}, 2\mu^{(1)} - \alpha^{(1)}\rangle = |us\rangle + |su\rangle.$$

Thus we get the weight vector state

$$|2\mu^{(1)}, 2\mu^{(1)} - \alpha^{(1)}\rangle = \frac{1}{\sqrt{2}}(|us\rangle + |su\rangle) \equiv |\boxed{u\,s}\rangle.$$

Further operating the same lowering operator, we can show that

$$|2\mu^{(1)}, 2\mu^{(1)} - 2\alpha^{(1)}\rangle = |\boxed{s\ s}\rangle \equiv |ss\rangle.$$

Remember that $J_-^{(\alpha^{(2)})}|2\mu^{(1)}, 2\mu^{(1)}\rangle = 0$ as $\alpha^{(2)} \cdot \mu^{(1)} = 0$. However, we can operate the lowering operator associated with the positive root $\alpha^{(3)} = \alpha^{(1)} + \alpha^{(2)}$ to derive the following two states:

$$|2\mu^{(1)}, 2\mu^{(1)} - \alpha^{(1)} - \alpha^{(2)}\rangle = \frac{1}{\sqrt{2}}(|ud\rangle + |du\rangle) \equiv |\boxed{u\ d}\rangle;$$

$$|2\mu^{(1)}, 2\mu^{(1)} - 2\alpha^{(1)} - 2\alpha^{(2)}\rangle = |\boxed{d\ d}\rangle \equiv |dd\rangle.$$

On the state $|\boxed{d\ d}\rangle$, we cannot apply $J_-^{(\alpha^{(2)})}$ as the dot product of the root vector with the weight vector being $2\alpha^{(2)} \cdot (2\mu^{(1)} - 2\alpha^{(1)} - 2\alpha^{(2)})/|\alpha^{(2)}|^2 = (q - p) = -2$. Hence, we will try raising the operator twice on the state. Applying once gives

$$|2\mu^{(1)}, 2\mu^{(1)} - 2\alpha^{(1)} - \alpha^{(2)}\rangle = \frac{1}{\sqrt{2}}(|ds\rangle + |sd\rangle) \equiv |\boxed{d\ s}\rangle,$$

and applying the raising operator on the $|\boxed{d\ s}\rangle$ will give the state $|\boxed{s\ s}\rangle$ already obtained. Thus we have derived the six weight vector states and presented the CG coefficient decomposition of the di-quark bound state by applying the $\mathfrak{su}(3)$ raising and lowering operators as briefly mentioned in Section 6.2.2. In the similar way, we can write the highest weight vector state $|3\mu^{(1)}, 3\mu^{(1)}\rangle \equiv \boxed{\phantom{uuu}} = |uuu\rangle$ and determine the weight states corresponding to the remaining nine baryons in the decimet diagram using the CG construction method.

## 6.4.2 Antiparticles

Antiparticles of quarks are called antiquarks. Just as for particles, we can have decimet diagrams and octet diagrams for antibaryons. The values of $I_z, Y, Q, S, B$ of antiparticles are opposite in sign with respect to their corresponding particles. We will now describe the Young diagram for the antiparticle multiplet of the $SU(N)$ group.

Suppose a particle multiplet is $(a_1, a_2, \ldots a_{N-1})$ where $a_1, a_2, \ldots a_{N-1}$ are the number of single box, number of double vertical boxes, $\cdots$ number of columns of $N - 1$ vertical boxes in the corresponding Young diagram presentation. Then the antiparticle multiplet will be given by $(a_{N-1}, a_{N-2}; \ldots a_2, a_1)$. For example, $SU(3)$ decimet baryon multiplet $\boxed{\phantom{ab}}$ denoted by $(a_1 = 3, a_2 = 0)$. The corresponding antibaryon decimet is $(a_2 = 0, a_1 = 3)$ whose Young diagram is $\boxed{\phantom{ab}}$. Similarly,

$SU(3)$ antiquark representation is (0, 1) whose Young diagram is ▢. The highest weight vector of the antiquark state is $\mu^{(2)}$. If we change the signs of $x_i$'s and $y_i$'s of the points $(x_i, y_i)$ in the $SU(3)$ fundamental representation weight diagram in the $H_1$, $H_2$ plane (see Figure 5.7.2) denoting $u, d, s$ quarks, we get the weight diagram of $SU(3)$ antiquarks. The readers can verify that the highest weight vector of the antiquark multiplet is $\mu^{(2)}$ (as determined in Example 48). Using these Young diagrams of the antiparticle multiplet, we can study mesons which are bound states of quarks and antiquarks.

### Mesons

In group theory language, we say that the $SU(3)$ mesons belong to irreducible representations obtained from tensor products of quark and antiquark representation:

$$\square \otimes \begin{array}{c}\square\\\square\end{array} = \begin{array}{c}\square\\\square\\\square\end{array} \oplus \begin{array}{c}\square\square\\\square\end{array} \tag{6.4.2}$$

where the first irreducible representation (vertical 3 boxes) is a trivial $SU(3)$ representation whose isospin $\vec{I}, I_z$, hypercharge $Y$ are zero. The second irreducible representation whose dimension is 8 is called octet meson multiplet. Notice that the meson bound state multiplet is its own antimeson multiplet. That is, some mesons are their own antimesons and some mesons get mapped to another set of mesons within the octet multiplet.

During the 1974 - 1984 period, new mesons and baryons were discovered which required the introduction of more flavors for quarks besides $u$, $d$, $s$. As of today, there are actually six flavors of quarks indirectly observed through scattering experiments. They are called up ($u$), down (d), strange ($s$), charm ($c$), beauty ($b$) and top ($t$) quarks. We could extend Gell-Mann's quark model proposal by including these flavors as well. That is, the Lie group will be $SU(N)$ where $N$ denotes the number of quark flavors. We have given exercise problems to use the Young diagram tensor product tools to understand such new baryons and meson bound states.

We will now discuss in the following subsection as to how the $SU(3)$ irreducible representations like decimet and octet will break if there is a perturbation that breaks the symmetry to a lower rank group. In Chapter 4, we discussed (in the discrete group context) how some degenerate energy levels split into non-degenerate states. In similar fashion, we would like to understand how the ten-dimensional decimet belonging to $SU(3)$ breaks to a lower rank group symmetry by adding perturbations.

### 6.4.3 Symmetry breaking from $SU(3) \to SU(2)$

In the absence of perturbation, the system described by the Hamiltonian $H_0$ possesses $SU(3)$ symmetry if and only if the Hamiltonian commutes with all the eight generators of $\mathfrak{su}(3)$. Suppose we add a perturbation term to the Hamiltonian $H_1 = k\sum_{i=1}^{3}\lambda_i^2$ where $k$ is a constant ($|k| < 1$) and $\lambda_{\{i=1,2,3\}}$ are the first three Gell-Mann matrices, then the total Hamiltonian

$$[H = H_0 + H_1, \lambda_i] = 0,$$

where $i = 1, 2, 3$. Note that the first three Gell-Mann matrices satisfy $\mathfrak{su}(2)$ algebra. Suppose the system is the quark system described by defining representation, we see that

$$(\lambda_1 + i\lambda_2)|\Lambda, \mu_1 \equiv u\rangle \propto |\Lambda, \mu_2 \equiv d\rangle; (\lambda_1 - i\lambda_2)|\Lambda, \mu_2 \equiv d\rangle \propto |\Lambda, \mu_1 \equiv u\rangle,$$

whereas the state $|\Lambda, \mu_3\rangle \equiv |s\rangle$ remains invariant under the action of $\lambda_{\{i=1,2,3\}}$ which means that the state transforms as a one-dimensional representation of $SU(2)$. That is, the three-dimensional defining representation under the perturbation $H_1$ breaks as

$$3 \to 2 + 1.$$

The three-dimensional multiplet becomes

$$(u\ d\ s) = (u\ d) \oplus (s).$$

**Example 60.** Suppose the perturbation breaks the $SU(3)$ symmetry to $SU(2)$ such that s quarks behave like a trivial (singlet) representation of $SU(2)$. How will the $SU(3)$ baryons in the decimet multiplet break under such a perturbation?

Recall, that in the decimet weight diagram $\Delta$ particles are made of non-strange quarks. Hence they transform like $\boxed{\phantom{x}\phantom{x}\phantom{x}} \in SU(2)$ which is four-dimensional. The baryons $\Sigma^*$ have one strange quark and hence will transform as $\boxed{\phantom{x}\phantom{x}} \in SU(2)$ as one of the boxes with s quark behaves trivial. Similarly $\Xi^*$ has two s quarks which will transform as $\boxed{\phantom{x}} \in SU(2)$. The $\Omega^-$ has three s quarks and hence will transform as a singlet representation. Therefore

$$10 \to 4 + 3 + 2 + 1$$

is the way in which decimet breaks under the perturbation which reduces the $SU(3)$ symmetry to $SU(2)$ symmetry. □

So far, we have discussed applications in quantum mechanics, particle and nuclear physics in an elaborate fashion. Particularly, we have examined the tensor product of $SU(2)$, $SU(3)$ representations which is generalizable for $SU(N)$ groups and their relevance to selection rules and elementary particle multiplets. In the following section, we will discuss the non-compact group symmetry which forms the backbone of quantum theories in 3+1 dimensional flat space-time (also known as Minkowski space-time).

## 6.5 Non-compact Groups

We will discuss the non-compact group symmetries like Lorentz group, Poincare groups, conformal groups in this section.

### 6.5.1 The Lorentz group

The Lorentz group $SO(3,1)$ is a set of $\{\Lambda\}$ $4 \times 4$ matrices with real matrix elements and determinant $+1$ satisfying

$$\Lambda^T \eta \Lambda = \eta,$$

where $\eta$ is a $4 \times 4$ diagonal matrix (also known as Minkowski metric):

$$\eta = \begin{pmatrix} 1 & 0 & 0 & 0 \\ 0 & -1 & 0 & 0 \\ 0 & 0 & -1 & 0 \\ 0 & 0 & 0 & -1 \end{pmatrix}.$$

Using the above condition, we can deduce that the number of independent matrix elements is 6. That is, there must be six Lie algebra generators. These $4 \times 4$ matrices $\Lambda$ act on the four-dimensional space-time coordinates $x^\mu = (x^0, \vec{x}) = (ct, x, y, z)$. Alternatively, the group is a set of transformations

$$x^\mu \to x'^\mu = \Lambda^\mu_\nu x^\nu; \equiv \begin{pmatrix} ct' \\ x' \\ y' \\ z' \end{pmatrix} = \Lambda^\mu_\nu \begin{pmatrix} ct \\ x \\ y \\ z \end{pmatrix}$$

such that

$$(cdt \;\; dx \;\; dy \;\; dz) \begin{pmatrix} 1 & 0 & 0 & 0 \\ 0 & -1 & 0 & 0 \\ 0 & 0 & -1 & 0 \\ 0 & 0 & 0 & -1 \end{pmatrix} \begin{pmatrix} cdt \\ dx \\ dy \\ dz \end{pmatrix} =$$

$$(cdt' \;\; dx' \;\; dy' \;\; dz') \begin{pmatrix} 1 & 0 & 0 & 0 \\ 0 & -1 & 0 & 0 \\ 0 & 0 & -1 & 0 \\ 0 & 0 & 0 & -1 \end{pmatrix} \begin{pmatrix} cdt' \\ dx' \\ dy' \\ dz' \end{pmatrix}.$$

The familiar rotation group $SO(3)$ in the three-dimensional space $(x, y, z)$ is a subgroup $SO(3,1)$. Besides the rotation operation, we can also consider the following operation called Lorentz transformation by boosting along the $x$-direction:

$$t' = \frac{t - \frac{v_x x}{c^2}}{\sqrt{1 - \frac{v_x^2}{c^2}}}; \quad x' = \frac{x - v_x t}{\sqrt{1 - \frac{v_x^2}{c^2}}}; \quad y' = y; \quad z' = z, \qquad (6.5.1)$$

where $v_x$ is the x-component boost velocity which serves as one of the parameters of the Lorentz group. Thus, the parameters corresponding to three rotations $\theta_x, \theta_y, \theta_z$ about the $x$, $y$, $z$ axes plus the three boost parameters $v_x, v_y, v_z$ corresponding to Lorentz transformation along $x$, $y$, $z$ directions combine to give six independent parameters. We can determine the matrix form of the corresponding generators associated with the infinitesimal transformations.

**Generators of the Lorentz group**

The generators of rotation for the Lorentz group along the three coordinate axes $(x, y, z)$ are the same as the generators of the $SO(3)$ group that forms a closed algebra of angular momentum operators $L$ as discussed in Chapter 5. Instead of our earlier notion of rotation about axis, we will state the rotation operation in a plane so that the we can generalize the operation to four-dimensional Minkowski space-time as well as rotations in any higher dimensional space. For example, consider an infinitesimal rotation about z-axis. We will denote the rotation parameter as $\delta\phi_{xy}$ (as a rotation about z-axis which mixes $x$ and $y$ components):

$$R(\delta\theta_z) \equiv R(\delta\phi_{xy}), \qquad (6.5.2)$$

$$R(\delta\phi_{xy}) = \begin{pmatrix} \cos\delta\phi_{xy} & -\sin\delta\phi_{xy} & 0 \\ \sin\delta\phi_{xy} & \cos\delta\phi_{xy} & 0 \\ 0 & 0 & 1 \end{pmatrix} = \mathbb{I} + \begin{pmatrix} 0 & -1 & 0 \\ 1 & 0 & 0 \\ 0 & 0 & 0 \end{pmatrix} \delta\phi_{xy} = \mathbb{I} + L_{xy}\delta\phi_{xy},$$

where $L_{xy} = \begin{pmatrix} 0 & -1 & 0 \\ 1 & 0 & 0 \\ 0 & 0 & 0 \end{pmatrix}$ is the generator for rotation in the x-y plane and $\delta\phi_{xy}$ is the corresponding parameter for this transformation. The matrix form of these generators must be $4 \times 4$ in the four-dimensional Minkowski space-time. That is we should add an extra first row and a first column associated with the time coordinate with 0 as entries:

$$L_{xy} = \begin{pmatrix} 0 & 0 & 0 & 0 \\ 0 & 0 & -1 & 0 \\ 0 & 1 & 0 & 0 \\ 0 & 0 & 0 & 0 \end{pmatrix}. \qquad (6.5.3)$$

Similarly, the generators for rotation in the y-z plane (about the x-axis) and z-x plane (about the y-axis) respectively are

$$L_{yz} = \begin{pmatrix} 0 & 0 & 0 & 0 \\ 0 & 0 & 0 & 0 \\ 0 & 0 & 0 & -1 \\ 0 & 0 & 1 & 0 \end{pmatrix}; \quad L_{zx} = \begin{pmatrix} 0 & 0 & 0 & 0 \\ 0 & 0 & 0 & 1 \\ 0 & 0 & 0 & 0 \\ 0 & -1 & 0 & 0 \end{pmatrix} \quad (6.5.4)$$

These three rotation generators in Equations 6.5.3 and 6.5.4 satisfy the commutation relation

$$[\hat{L}_i, \hat{L}_j] = \epsilon_{ijk} \hat{L}_k, \quad (6.5.5)$$

where $\hat{L}_p = \frac{1}{2} \epsilon_{pqr} L_{qr}$, with the $p$, $q$, $r \in (x, y, z)$ and $\epsilon_{pqr}$ is the Levi–Civita antisymmetric tensor of rank 3 where

$$\epsilon_{xyz} = \epsilon_{yzx} = \epsilon_{zxy} = -\epsilon_{yxz} = -\epsilon_{xzy} = -\epsilon_{zyx} = +1,$$

and zero otherwise. Similar to the matrix form of rotation generators, we will determine the matrix form of boost generators by rewriting the Lorentz transformation (6.5.1) operation using the notation $\beta = \frac{v_x}{c}; \gamma = \frac{1}{\sqrt{1-\beta^2}}$

$$\begin{pmatrix} ct' \\ x' \\ y' \\ z' \end{pmatrix} = \begin{pmatrix} \gamma & -\beta\gamma & 0 & 0 \\ -\beta\gamma & \gamma & 0 & 0 \\ 0 & 0 & 1 & 0 \\ 0 & 0 & 0 & 1 \end{pmatrix} \begin{pmatrix} ct \\ x \\ y \\ z \end{pmatrix}. \quad (6.5.6)$$

Implementing the following change of variables:

$$\gamma = \cosh \phi_{xt} \text{ and } \beta\gamma = \sinh \phi_{xt} \Rightarrow \beta = \tanh \phi_{xt}, \quad (6.5.7)$$

where the quantity $\phi_{xt}$ is known in literature as 'rapidity' which is related to velocity $v_x$ as $\phi_{xt} = \tanh^{-1} \beta$. The subscript $xt$ on the rapidity parameter keeps track of the boost along x-direction which mixes the $x$, $t$ coordinates. We can now determine the boost generator from the infinitesimal transformation:

$$\Lambda(\delta\phi_{xt}) = \begin{pmatrix} \gamma & -\beta\gamma & 0 & 0 \\ -\beta\gamma & \gamma & 0 & 0 \\ 0 & 0 & 1 & 0 \\ 0 & 0 & 0 & 1 \end{pmatrix} = \mathbb{I} + K_{xt}\delta\phi_{xt}, \quad (6.5.8)$$

where $K_{xt}$ is the generator for Lorentz boost along $x$-direction and $\delta\phi_{xt}$ is the parameter corresponding to the transformation

$$\Lambda(\delta\phi_{xt}) = \begin{pmatrix} \cosh\delta\phi_{xt} & -\sinh\delta\phi_{xt} & 0 & 0 \\ -\sinh\delta\phi_{xt} & \cosh\delta\phi_{xt} & 0 & 0 \\ 0 & 0 & 1 & 0 \\ 0 & 0 & 0 & 1 \end{pmatrix} = \mathbb{I} + \begin{pmatrix} 0 & -1 & 0 & 0 \\ -1 & 0 & 0 & 0 \\ 0 & 0 & 0 & 0 \\ 0 & 0 & 0 & 0 \end{pmatrix} \delta\phi_{xt}.$$

(6.5.9)

So, now, comparing Equations 6.5.8 and 6.5.9, we get the Lorentz boost generator along $x$-direction:

$$K_{xt} = \begin{pmatrix} 0 & -1 & 0 & 0 \\ -1 & 0 & 0 & 0 \\ 0 & 0 & 0 & 0 \\ 0 & 0 & 0 & 0 \end{pmatrix}. \tag{6.5.10}$$

Similarly, the generators for boost along the $y$-direction and $z$-direction respectively are

$$K_{yt} = \begin{pmatrix} 0 & 0 & -1 & 0 \\ 0 & 0 & 0 & 0 \\ -1 & 0 & 0 & 0 \\ 0 & 0 & 0 & 0 \end{pmatrix}; \quad K_{zt} = \begin{pmatrix} 0 & 0 & 0 & -1 \\ 0 & 0 & 0 & 0 \\ 0 & 0 & 0 & 0 \\ -1 & 0 & 0 & 0 \end{pmatrix}. \tag{6.5.11}$$

Rewriting $\hat{K}_i = K_{it}$ where $i$ denotes the $x$, $y$, $z$ coordinates, it is simple to verify that the matrices in Equations 6.5.10, 6.5.11, 6.5.3 and 6.5.4 satisfy the following commutation relations:

$$[\hat{K}_i, \hat{K}_j] = -\epsilon_{ijk}\hat{L}_k, \tag{6.5.12}$$

$$[\hat{L}_i, \hat{K}_j] = \epsilon_{ijk}\hat{K}_k. \tag{6.5.13}$$

So, the six generators of the Lorentz group ($L_{xy}$, $L_{yz}$, $L_{zx}$, $K_{xt}$, $K_{yt}$, $K_{zt}$) satisfy the three commutation relations given in Equations 6.5.5, 6.5.12 and 6.5.13. The Lie algebra of the Lorentz group is based on these three commutation relationships of the generators.

The rotation group is compact, i.e., the parameter space for rotation, consisting of angles, is bound between 0 and $2\pi (0 \leq \phi_{xy}, \phi_{yz}, \phi_{zx} \leq 2\pi)$; while the boost operation is non-compact, i.e., the parameter space for boost transformations ranges from $-\infty$ to $\infty (-\infty \leq \phi_{xt}, \phi_{yt}, \phi_{zt} \leq \infty)$, although the boost velocity is constrained by the special theory of relativity by $-c \leq v_x, v_y, v_z \leq c$.

Thus the Lorentz group $SO(3,1)$ is a non-compact group whose Lie algebra involves three rotation generators as well as three boost generators.

### 6.5.2 Poincare group

Besides the Lorentz group symmetry, physical systems can respect space translational and the time translational symmetry. The generators of the space-time translation in relativistic notation is four momentum vector: $p^\mu = (E/c, \vec{p})$. Recall that the linear momentum $\vec{p}$ corresponds to space translation generator and energy $E$ corresponds to the time translational generator. The readers can extend the algebra by including $E, p_x, p_y, p_z$ along with the six Lorentz generators and thereby construct Poincare algebra.

### 6.5.3 Scale invariance

All physical systems near their phase transition point are called critical systems. For example, near liquid-vapor phase transition, water molecules of all possible sizes will be present. In other words, the systems at criticality possess scale symmetry: $x'^\mu = \exp(\alpha) x^\mu$.

**Example 61.** Determine the generator $D$ for the scale transformation:

$$x'^\mu = \exp(\alpha) x^\mu.$$

For infinitesimal scale transformation any function

$$f(x'^\mu) = f(x^\mu + \alpha x^\mu) = f(x^\mu) + \alpha x^\mu \frac{\partial}{\partial x^\mu} f(x^\mu),$$

implying that the generator is

$$D = x^\mu \frac{\partial}{\partial x^\mu}.$$

### 6.5.4 Conformal group

In addition to the scaling symmetry and Poincare symmetry, there is a symmetry called special conformal transformation which is defined as follows:

$$x^\mu \to x'^\mu = \frac{x^\mu}{|x|^2} \to x''^\mu = x'^\mu + b^\mu \to x'''^\mu = \frac{x''^\mu}{|x''|^2}.$$

In words, it is an inversion followed by a translation by vector $b^\mu$ and again another inversion. There will be four generators associated with the four parameters. The algebra of $E, p_x, p_y, p_z, \hat{L}_i, \hat{K}_i, D$ plus the four special conformal transformation generators form an algebra called conformal algebra. These symmetries find applications in quantum field theories called conformal field theories, the conformal group symmetries. Interested readers can pursue these areas for research.

## 6.6 Dynamical Symmetry in Hydrogen Atom

The resemblance between the Kepler problem of planetary motion and the charged particle in Coulombic potential motivates us to understand symmetries possessed by such systems. We can formally write the Hamiltonian as

$$H = \frac{\vec{p}^2}{2\mu} - \frac{\kappa}{r}, \tag{6.6.1}$$

where $\mu$ is the reduced mass $[Mm/(M+m)]$, $\kappa = GMm$ (for gravitational potential) and $\kappa = Ze^2$ (for Coulombic potential where $Z$ is the atomic number). These systems possess rotational symmetry and hence the angular momentum $\vec{L} = \vec{r} \times \vec{p}$ (generators of rotations in three-dimensional space) is a conserved quantity. Further, for such a central force potential satisfying the inverse-square law, there is one another conserved quantity known as the *Runge–Lenz* vector:

$$\vec{M} = \frac{\vec{p} \times \vec{L}}{\mu} - \frac{\kappa}{r}\hat{r}. \tag{6.6.2}$$

Unlike the geometrical interpretation for rotation generators, we do not have a geometrical description for the Runge–Lenz vector. Hence we say that $\vec{M}$ represents generators of dynamical (hidden) symmetry. Classically, $\vec{M}$ satisfies the following relations:

$$\vec{L} \cdot \vec{M} = 0 \; ; \; \vec{M}^2 = \frac{2H}{\mu}\vec{L}^2 + \kappa^2, \tag{6.6.3}$$

where $H$ is the Hamltonian of the system.

The quantum mechanical operator $\hat{M}$ of the classical Runge–Lenz vector $\vec{M}$ will be defined as follows:

$$\hat{M} = \frac{1}{2\mu}(\hat{p} \times \hat{L} - \hat{L} \times \hat{p}) - \frac{\kappa}{r}\hat{r}, \tag{6.6.4}$$

as we require Hermitian operators corresponding to observables in quantum mechanics. This operator (6.6.4) commutes with the Hamiltonian, $[\hat{M}, H] = 0$, and hence is a conserved quantity. The relations in Equation 6.6.3 become

$$\hat{L} \cdot \hat{M} = \hat{M} \cdot \hat{L} = 0 \; ; \; \hat{M}^2 = \frac{2H}{\mu}(\hat{L}^2 + \hbar^2) + \kappa^2. \tag{6.6.5}$$

We will now present the Lie algebra determined by working out the commutator brackets amongst the components of $\hat{M}$ as well as the commutator bracket between $\hat{M}$ and $\hat{L}$.

## 6.6.1 Lie algebra symmetry

The commutation relations satisfied by the components $M_x$, $M_y$, $M_z$ are as follows:

$$[M_i, L_i] = 0, \tag{6.6.6}$$

$$[M_i, L_j] = i\hbar \epsilon_{ijk} M_k, \tag{6.6.7}$$

$$[M_i, M_j] = -\frac{2i\hbar}{\mu} \epsilon_{ijk} H L_k, \tag{6.6.8}$$

where $i, j, k = x, y, z = 1, 2, 3$.

Replacing the Hamiltonian operator $H$ by its corresponding energy eigenvalue $E$, and rescaling $M$ as

$$\hat{M} \to \hat{M}' = a\hat{M},$$

the commutation relation (6.6.8) becomes

$$\frac{1}{a^2}[M'_x, M'_y] = \left(-\frac{2i\hbar}{\mu}E\right) L_z. \tag{6.6.9}$$

So, the $M_i$'s and $L_i$'s as generators constitute a closed $SO(4)$ algebra if

$$a^2 \left(\frac{-2}{\mu}E\right) = 1 \Rightarrow a = \left(-\frac{\mu}{2E}\right)^{1/2}, \tag{6.6.10}$$

$$\hat{M}' = \left(-\frac{\mu}{2E}\right)^{1/2} \hat{M}. \tag{6.6.11}$$

Using this rescaling factor, Equation 6.6.8 becomes

$$[M'_i, M'_j] = i\hbar \epsilon_{ijk} L_k. \tag{6.6.12}$$

Comparing with the exercise problem (14), these generators can be mapped to $SO(4)$ group generators as follows:

$$\hat{M}'_i \equiv L_{i4}; \quad \hat{L}_i \equiv \frac{1}{2}\epsilon_{ijk} L_{jk},$$

In other words, we have to add a fourth fictitious coordinate $\omega$ such that the following operation defines $SO(4)$ group operation:

$$\begin{pmatrix} x' \\ y' \\ z' \\ \omega' \end{pmatrix} = e^{\Sigma_{i<j} \theta_{ij} L_{ij}} \begin{pmatrix} x \\ y \\ z \\ \omega \end{pmatrix}. \qquad (6.6.13)$$

Thus the $SO(4)$ group symmetry respected by the hydrogen atom Hamiltonian has no geometric interpretation in the three-dimensional physical space. Hence the $SO(4)$ symmetry is referred to as the dynamical symmetry respected by the hydrogen atom.

In quantum mechanics textbooks, the hydrogen atom energy spectrum $E_n = -13.6 eV/n^2$ is determined by solving the time independent Schrödinger equation. In the following subsection, our aim is to highlight the power of SO(4) dynamical symmetry. By exploiting this dynamical symmetry, we obtain the energy levels of the hydrogen atom ($1/n^2$ dependence of energy spectrum) without solving the Schrödinger equation.

### 6.6.2 Energy levels of hydrogen atom

The SO(4) group (generated by operators $\hat{L}$ and $\hat{M}'$) can be decomposed into two independent SU(2) groups. That is,

$$SO(4) \equiv SU(2) \times SU(2).$$

The generators of the two $SU(2)$ groups are

$$\hat{I} = \frac{1}{2}(\hat{L} + \hat{M}'); \quad K = \frac{1}{2}(\hat{L} - \hat{M}'), \qquad (6.6.14)$$

which satisfies the following commutation relations:

$$[I_i, I_j] = i\hbar \epsilon_{ijk} I_k,$$

$$[K_i, K_j] = i\hbar \epsilon_{ijk} K_k,$$

$$[\hat{I}, \hat{K}] = 0.$$

We observe that the algebra of the operators involving $\hat{I}$ and $\hat{K}$ form two independent $su(2)$ algebra which commutes with the Hamiltonian:

$$[\hat{I}, H] = 0; \quad [\hat{K}, H] = 0.$$

Hence we can construct states $|E, i, i_z, k, k_z\rangle$ which are simultaneous eigenstates of $\hat{I}_z, \hat{K}_z, H$ where E is the energy eigenvalue, $i, k$ denote highest weights and $i_z, k_z$ are the weights obtained by the operation of $\hat{I}_z, \hat{K}_z$ operators:

$$I_z|E, i, i_z, k, k_z\rangle = i_z\hbar|E, i, i_z, k, k_z\rangle,$$

$$K_z|E, i, i_z, k, k_z\rangle = k_z\hbar|E, i, i_z, k, k_z\rangle.$$

The highest weight can be obtained by the action of $\hat{I}.\hat{I} = I_z^2 + (I_+I_- + I_-I_+)$ and $\hat{K}.\hat{K} = K_z^2 + (K_+K_- + K_-K_+)$ on the states as follows:

$$\hat{I}.\hat{I}|E, i, i_z, k, k_z\rangle = i(i+1)\hbar^2|E, i, i_z, k, k_z\rangle,$$

$$\hat{K}.\hat{K}|E, i, i_z, k, k_z\rangle = k(k+1)\hbar^2|E, ; i, i_z, k, k_z\rangle \tag{6.6.15}$$

We now construct two new operators,

$$C_1 = \hat{I}.\hat{I} + \hat{K}.\hat{K} \quad ; \quad C_2 = \hat{I}.\hat{I} - \hat{K}.\hat{K}. \tag{6.6.16}$$

Using Equations 6.6.5 and 6.6.14, we can confirm that

$$C_2 = \hat{I}.\hat{I} - \hat{K}.\hat{K} = 0 \Rightarrow \hat{I}.\hat{I} = \hat{K}.\hat{K}. \tag{6.6.17}$$

The above condition forces that the highest weights $i$, $k$ cannot be independent but equal. Further incorporating this condition, the action of $C_1$ on the states will be

$$C_1|E, i, i_z, k, k_z\rangle = 2i(i+1)\hbar^2|E, i, i_z, k, k_z\rangle. \tag{6.6.18}$$

Substituting Equations 6.6.5, 6.6.11, 6.6.14, and 6.6.16, the form of the operator $C_1$ is:

$$C_1 = \frac{1}{2}(\hat{L}^2 + \hat{M}'^2) = -\frac{\hbar^2}{2} - \frac{\mu\kappa^2}{4H}. \tag{6.6.19}$$

Comparing the above equation with Equation 6.6.18, we deduce:

$$C_1|E, i, i_z, k, k_z\rangle = 2i(i+1)\hbar^2|E, i, i_z, k, k_z\rangle$$

$$= \left(-\frac{\hbar^2}{2} - \frac{\mu\kappa^2}{4E}\right)|E, i, i_z, k, k_z\rangle.$$

which implies

$$2i(i+1)\hbar^2 = -\frac{\hbar^2}{2} - \frac{\mu\kappa^2}{4E}, \tag{6.6.20}$$

indicating that the energy eigenvalues are

$$E_i = \frac{-\mu\kappa^2}{2\hbar^2(2i+1)^2}, \tag{6.6.21}$$

where we have put a subscript on the energy $E$ to keep track of the highest weight $i$. Recall that the highest weights of $\mathfrak{su}(2)$ algebra could be half-odd integers or integers. Hence,

$$2i + 1 = n,$$

where $n$ is always an integer. Substituting the value of $\kappa$ for hydrogen atom, the energy eigenvalues (6.6.21) in terms of $n$ are exactly same as that obtained from the Schrödinger equation:

$$E_n = -\frac{\mu e^4}{2\hbar^2 n^2}, \tag{6.6.22}$$

indicating that the integer $n$ can be referred to as the principal quantum number. The energy levels in the hydrogen atom are degenerate with degeneracy of $n^2$. Interestingly, using the $SU(2) \times SU(2)$ symmetry arguments we can reproduce the degeneracy of the energy levels. The physical orbital angular momentum operator $\hat{L} = \hat{I} + \hat{K}$ involves addition of two $\mathfrak{su}(2)$ generators.

Applying the CG construction method, we can obtain eigenstates $|\ell\ell_z\rangle$ of $\hat{L}_z$ from the simultaneous eigenstates of $I_z, K_z$ In fact, we can show that the the allowed highest weight $\{\ell\}$ corresponding to $\hat{L}$ generators will range from $i+k, i+k-1, \ldots |i-k|$. Substituting $i = k$ (6.6.17) will give $\ell = 2i, 2i-1, \ldots 0$. In terms of $n$, the range of $\ell = 0, 1, \ldots n-1$. For each $\ell$, the states $|\ell\ell_z\rangle$ allow $2\ell + 1$ states. The energy eigenvalues (6.6.22) are independent of both $\ell$ and $\ell_z$ accounting for degeneracy as

$$\text{degeneracy} = \sum_{\ell=0}^{n-1}(2\ell+1) = n^2.$$

## Exercises

1. Quadrapole moment tensor $Q(2, q)$ is derivable from the tensor product of $\vec{r} \otimes \vec{r} = \oplus_{q=-2}^{2} Q(2, q)$. Show that $Q(2,0) = 2z^2 - r^2$ and $Q(2,2) - Q(2,-2) = x^2 - y^2$.

2. Consider two spin 1/2 particles (proton and neutron) described by the Hamiltonian $H = k\hat{s}_n.\hat{s}_p$ where $k$ is a constant. Find the symmetry possessed by the system and the corresponding conserved quantity. Determine the energy eigenstates and eigenvalues.

3. If $S(k,q)$ and $T(k,q)$ are two irreducible tensor operators of rank $k$, prove that $\omega = \sum_{q=-k}^{k}(-1)^q T(k,q) S(k,-q)$ is a scalar operator.

4. A deuteron has spin 1. Use the Wigner–Eckart theorem to find the ratios of the expectation values of the electric quadrupole moment operator $Q(2,0)$ for the three orientations of deuteron ($m = 0, +1, -1$).

5. Using the Clebsch–Gordan coefficient,

$$\langle j_1 = j,\ j_2 = 2,\ m_1 = j,\ m_2 = 0 | J = j,\ m = j \rangle = \sqrt{\frac{j(2j-1)}{(j+1)(2j+3)}},$$

verify the following statement for quadrupole moment tensor: The static $2^k$ pole moment of a charge distribution has zero expectation value in any state with angular momentum $j < (k/2)$.

6. Show that the strong decay process $\rho^0 \to \pi^0 \pi^0$ is forbidden.

7. The highest weight vector state for decimet irreducible representation belonging to $SU(3)$ group is

$$|3\mu^{(1)}, 3\mu^{(1)}\rangle \equiv |\,\boxed{\phantom{xx}\,\phantom{xx}\,\phantom{xx}}\,\rangle = |uuu\rangle.$$

Determine the weight vector states corresponding to the remaining nine baryons in the decimet diagram using the CG construction method on the tensor product of three fundamental quarks $\boxed{\phantom{x}} \in SU(3)$.

8. Suppose an irreducible representation $\boxed{\phantom{xx}}\phantom{x}\in SU(3)$ denotes a hypothetical particle multiplet. What will be the Young diagram depicting the corresponding antiparticle multiplet?

9. Suppose we rewrite the Lorentz group $SO(3,1)$ generators as $M_i = \hat{L}_i + i\hat{K}_i$ and $N_i = \hat{L}_i - i\hat{K}_i$. Note that the $M_i$, $N_i$ are complex conjugates of each other. Write the algebra satisfied by these $M_i$'s and $N_i$'s.

10. Suppose we are given a six-dimensional multiplet belonging to a irreducible representation of a unitary group of rank less than 3. How do we check whether the six states $|i\rangle$ belong to spin 5/2 of $SU(2)$ ($\boxed{\phantom{xx}\,\phantom{xx}\,\phantom{xx}\,\phantom{xx}\,\phantom{xx}}$) or symmetric tensor of rank 2 representation $\boxed{\phantom{xx}\,\phantom{xx}}$ of $SU(3)$? Explain.

11. Assume quarks occur in four flavors. For this case, determine the irreducible representations and their dimensions for baryons and mesons.

12. For quarks belonging to the fundamental representation of $SU(N)$, determine the dimensions of the meson multiplets.

13. Determine the irreducible representations and their dimensions for di-baryon bound states obtained from baryons belonging to the octet multiplet of $SU(3)$.

14. $SO(4)$ denotes the group of orthogonal matrices in four-dimensional space with determinant 1. (i) Using the orthogonality property, show that the number of independent parameters is 6. (ii) We know that $\phi_k (k = 1, 2, 3)$ denotes rotation

about the $k$-axis in three dimensions. Similarly, we can give geometrical meaning by choosing the six parameters as $\phi_{ij}(i > j,\ i = 1, 2, 3, 4)$ to denote rotation in the $i$-$j$ plane. Let $L_{ij} = r_i p_j - r_j p_i$ be the generators of rotation about $i$-$j$ plane. Show that the generators obey a closed algebra.

# Appendix A

# Maschke's Theorem

Let $V$ be a finite dimensional vector space (real or complex) and $T$ a linear operator on $V$. The kernel of the transformation $T$ is the subspace $K$ of $V$ on which $T$ vanishes. The image of $T$ is also a subspace of $V$, let this subspace be $R$. One seeks the condition on the operator $T$ so that any given vector $v \in V$ can be expressed uniquely as $v = k + r$ for some $k \in K$ and $r \in R$. Notice that $v = (I - T)v + Tv$ and $Tv \in R$. If this representation of $v$ is unique, then $(I - T)v \in K$. It follows that $T(I - T)v = 0$ for all $v \in V$ and one has the condition

$$T^2 = T \tag{A.0.1}$$

if $V = K \oplus R$. Conversely, suppose the above condition holds for $T$. If $u \in K \cap R$, then $Tu = 0$ and also there exists a $w \in V$ such that $Tw = u$. Then

$$u = Tw = TTw = Tu = 0.$$

It also follows from here that $v = (I - T)v + Tv$ is the desired unique representation of any given $v \in V$, i.e., $V = K \oplus R$. Any operator $T$ that satisfies the condition A.0.1 is called a projection operator. The reader can easily construct an operator on $\mathbb{R}^2$ which has a non-trivial kernel but is not a projection operator.

Now let $\Gamma_1$ and $\Gamma_2$ be representations of a group $G$ on vector spaces $V$ and $W$ respectively. A linear transformation $T$ from $V$ to $W$ is said to be *G-linear* if

$$T \circ \Gamma_1(g) = \Gamma_2(g) \circ T$$

for all $g \in G$. $K$ be the kernel of $T$ and $L = T(V)$ be the subspace of $W$. If $l \in L$, then there exists a $v \in V$ such that $l = Tv$. For any $g \in G$, $\Gamma_2(g)l = \Gamma_2(g) \circ Tv = T \circ \Gamma_1(g)v$, i.e., $\Gamma_2(g)l \in L$. Hence, $\Gamma_2$ is a subrepresentation of $G$ on $L$. A similar argument shows that $\Gamma_1$ is a subrepresentation of $G$ on $K$.

THEOREM. *Let $\Gamma$ be a representation of $G$ on a finite dimensional vector space $V$ ( Real or Complex). If $\Gamma$ is a subrepresentation on a subspace $L$ of $V$, then $\Gamma$ is a subrepresentation on another subspace $K$ of $V$ such that $V = K \oplus L$.*

PROOF. A basis of $L$ in $V$ can always be extended to a basis of $V$. Then the subspace $K'$ spanned by basis vectors required for the extension is such that $V = K' \oplus L$. For every $v \in V$, one has $v = k + l$ uniquely for some $k \in K'$ and $l \in L$. Define the projection operator $P$ on $V$ such that $Pv = l$. Then $K'$ is the kernel of $P$ and $L$ is the image of $P$. If $P$ was G-linear, then the theorem follows from the discussion prior to the statement of the theorem. However $P$ need not be G-linear. Consider the operator $T$ defined as

$$T = \frac{1}{|G|} \sum_{g \in G} \Gamma(g^{-1}) P \Gamma(g).$$

By construction, $T$ is G-linear (see also the discussion of Equation B.0.2). Also, for every $l \in L$, because $L$ is invariant under $\Gamma(g)$ for all $g \in G$ and $P$ is projection onto $L$, $Tl = l$. This shows that $T$ a map of $V$ into $V$ whose image is the subspace $L$. For these same reasons, $T^2 = T$ and it follows that $T$ is a projection. Let the kernel of $T$ be $K$, so that one has by the property of projection operator $T$

$$V = K \oplus L$$

and $\Gamma$ is a subrepresentation of $G$ over the subspace $K$. □

The Maschke's theorem states that every representation of a finite group over a finite dimensional vector space (real or complex) is completely reducible. This is a direct consequence of the above result.

# Appendix B

# Schur's Lemma

Schur's Lemma is stated and proved in the following. Some of its consequences, as regards irreducible representations and their characters are also developed. It is assumed that the vector spaces are finite dimensional over real or complex numbers and the groups are finite. The reader should be familiar with the notation and terminology introduced in Sections 3.1 to 3.3.

LEMMA. *(Schur) Let $\Gamma_1$ and $\Gamma_2$ be irreducible representations of a group G over vector spaces $V_1$ and $V_2$. If T is a linear transformation of $V_1$ into $V_2$ such that*

$$T\Gamma_1(g) = \Gamma_2(g)T \tag{B.0.1}$$

*for all g in G, then either T is an isomorphism of spaces $V_1$ and $V_2$ or the kernel of T is $V_1$. In the case T is an isomorphism with $\Gamma_1(g) = \Gamma_2(g)$ for all g in G, T is an scaling transformation.*

PROOF. Let the image of $T$ in $V_2$ be $L$. If $L$ is a non-trivial proper subspace of $V_2$, then the condition implies that $\Gamma_2$ is a representation on $L$. This is not possible since $\Gamma_2$ is irreducible over $V_2$. Then $L$ is either the null vector space or all of $V_2$. Similarly, the kernel $K$ of $T$ is either the null vector space or all of $V_1$.

In the case $K$ is the null vector space and $L = V_2$, $T$ is evidently an invertible linear transformation satisfying $T\Gamma_1(g)T^{-1} = \Gamma_2(g)$ for all $g$ in $G$, i.e., $\Gamma_1$ and $\Gamma_2$ are equivalent. $T$ may now be regarded as an invertible linear operator on the vector space $V(=V_1)$. There then exists a scalar $\lambda \neq 0$ and a vector $v \neq 0$ in $V$ such that

$$Tv = \lambda v.$$

Assuming $\Gamma_1 = \Gamma_2 = \Gamma$, the linear span $\mathcal{L}(\{\Gamma(g)v\}_{g \in G})$ is a representation subspace of $V$, and by irreducibility of $\Gamma$, $\mathcal{L}(\{\Gamma(g)v\}_{g \in G}) = V$. Any vector $v' \in V$, can be expressed in the form $v' = \sum_{g \in G} \alpha_g \Gamma(g) v$ for some choice of scalars $\alpha_g$. Then one has

$$Tv' = T\left(\sum_{g \in G} \alpha_g \Gamma(g) v\right) = \sum_{g \in G} \alpha_g T\Gamma(g) v = \sum_{g \in G} \alpha_g \Gamma(g) T v$$

$$\Rightarrow Tv' = \lambda v'. \qquad \square$$

In the general case, a transformation $T$ need not satisfy the condition (B.0.1). In order to successfully apply the Lemma, it is desirable to obtain a suitable transformation $T_S$ from a given $T$ so that $T_S$ fits the condition. Consider the following expression for $T_S$:

$$T_S = \frac{1}{|G|} \sum_{g \in G} \Gamma_2(g^{-1}) T \Gamma_1(g), \qquad (B.0.2)$$

where $\Gamma_1$ and $\Gamma_2$ are irreducible representations of the the group $G$ on vector spaces $V_1$ and $V_2$, while $T$ is a linear transformation from $V_1$ into $V_2$. Clearly, $T_S$ is also a linear transformation from $V_1$ into $V_2$. The form of $T_S$ suggests an averaging of $T$ over the group. For $h \in G$,

$$T_S \Gamma_1(h) = \frac{1}{|G|} \sum_{g \in G} \Gamma_2(g^{-1}) T \Gamma_1(g) \Gamma_1(h)$$

$$\Rightarrow T_S \Gamma_1(h) = \frac{1}{|G|} \sum_{g \in G} \Gamma_2(h) \Gamma_2(h^{-1}) \Gamma_2(g^{-1}) T \Gamma_1(g) \Gamma_1(h)$$

$$\Rightarrow T_S \Gamma_1(h) = \Gamma_2(h) \left[ \frac{1}{|G|} \sum_{g \in G} \Gamma_2((gh)^{-1}) T \Gamma_1(gh) \right]$$

$$\Rightarrow T_S \Gamma_1(h) = \Gamma_2(h) T_S$$

which is exactly the condition one needs for application of the Lemma. In Equation B.0.2, if $\Gamma_1 = \Gamma_2 = \Gamma$, then $T_S$ is a transformation of scale $\lambda$. If the degree of representation $\Gamma$ is $\ell_\Gamma$, then upon equating the trace of both sides of Equation B.0.2, one obtains

$$\lambda = \frac{tr(T)}{\ell_\Gamma}. \qquad (B.0.3)$$

In this case, $T_S$ is diagonal and each entry in the diagonal is equal to $\lambda$. Upon explicitly writing, say the element $(T_S)_{ii}$ from Equation B.0.2

$$\frac{\sum_{j=1}^{\ell_\Gamma} T_{jj}}{\ell_\Gamma} = \frac{1}{|G|} \sum_{g \in G} [\Gamma(g)]_{ik} T_{km} [\Gamma(g^{-1})]_{mi} \qquad (B.0.3)$$

and because the matrix $T$ is an arbitrary linear transformation, one can equate coefficients of $T_{km}$ on both sides of above to obtain

$$\sum_{g \in G} [\Gamma(g)]_{ik} [\Gamma(g^{-1})]_{ml} = \frac{|G|}{\ell_\Gamma} \delta_{il} \delta_{km}.$$

Similar calculation for $\Gamma_1 \neq \Gamma_2$ readily leads to

$$\sum_{g \in G} [\Gamma_1(g)]_{ik} [\Gamma_2(g^{-1})]_{ml} = 0.$$

If it is further assumed that the representations are unitary, and noting that in such a case $[\Gamma(g^{-1})]_{ml} = \overline{[\Gamma(g)]}_{lm}$, the above two expressions combined give Equation 3.3.5

$$\sum_{g \in G} [\Gamma(g)]_{ik} \overline{[\Theta(g)]}_{lm} = \frac{|G|}{\ell_\Gamma} \delta_{\Gamma\Theta} \delta_{il} \delta_{km}.$$

# Bibliography

[1] Georgi, Howard. 1999. *Lie Algebras in Particle Physics: From Isospin to Unified Theories (Frontiers in Physics)*. Boulder, Colorado: Westview Press.

[2] Hamermesh, M. 1962. *Group Theory and Its Applications to Physical Problems*. Reading, MA: Addison-Wesley.

[3] Joshi, A. W. 1973. *Elements of Group Theory for Physicists*. Delhi: New Age International (P) Ltd. Pub.

[4] Landau, L. D., and E. M. Lifshitz. 1965. *Quantum Mechanics: Non-Relativistic Theory*. Oxford: Pergamon Press. (See Chapter 13: Polyatomic Molecules, where molecular vibrations are discussed.)

[5] Ma, Zhong-Qi, and Xiao-Yan Gu. 2004. *Problems and Solutions in Group Theory for Physicists*. Singapore: World Scientific Publishing.

[6] O'Raifeartaigh, Lochlainn. 1986. *Group Structures of Gauge Theories*. N. York: Cambridge University Press.

[7] Ramond, Pierre. 2010. *Group Theory: A Physicist's Survey*. N. York: Cambridge University Press.

[8] Schiff, L. I. 1955. *Quantum Mechanics*. N. York: McGraw-Hill International Editions. (See Chapter 7: Symmetry in Quantum Mechanics.)

[9] Serre, Jean-Pierre. 1977. *Linear Representations of Finite Groups*. N. York: Springer-Verlag.

[10] Sury, B. 2004. *Group Theory: Selected Problems*. Hyderabad: Universities Press.

[11] Tung, Wu-Ki. 1985. *Group Theory in Physics*. Singapore: World Scientific Publishing.

[12] Zee, A. 2106. *Group Theory in a Nutshell for Physicists*. Princeton: Princeton University Press.

# Index

abelian group 2
adjoint 34
adjoint representation 98
alternating group 14
angular momentum algebra 93–94
   addition of angular momentum 116, 119
antisymmetric subspace 51

baryons 116
bilateral axis 23

Cartan
   matrix 113
   subalgebra 107
Cauchy–Schwartz inequality 33
Cayley's Theorem 10
character
   of the group element 38
   table 41
circle group $U(1)$ 87
Clebsch–Gordan coefficient 124
commutator 91
compact groups 95
conformal group 117, 143
conjugacy class 5

conjugate subgroup 5
coset 4
cyclic group 2

degenerate states 60
dihedral group 6
direct product 14
double cover 101
doubly connected space 95
dynamical symmetry 117, 144
Dynkin diagrams 109

eigenfrequencies 73
eigenvalue equation 32
electric dipole moment 66, 68
exceptional Lie algebra 110

factor group 5
Frobenius reciprocity theorem 57
fundamental weights 113

generating set 2
generators 2, 89
   constants of motion 117
gerade 41
Gram–Schmidt Orthogonalisation 33

Group 1
  continuous 88
  direct product of groups 14
  order of the group 2
  properties of the group 1
  representations 21

highest weight vector 108
homomorphism 8
  kernel of homormorphism 8
hydrogen atom 117, 144
  Runge–Lenz vector 117, 144
hypercharge 134

induced representation 54
inner product space 33
invariant subalgebra 102
irreducible representation 38
  of direct product groups 52
irreducible tensor 124
isomorphic groups 9
isospin 116, 131

Jacobi identity 92

Killing form 101
  degenerate Killing form 101
  nondegenerate Killing form 102
Klein-4 group 3

Lagrange theorem 4
Lagrangian 72
  Euler–Lagrange equation 72
level splitting 62
Lie algebra 92
  Cartan subalgebra 107
  complex 92
  real 92
  semi-simple Lie algebra 101
  simple Lie algebra 103
    compact simple Lie algebra 107
  structure constants 92

Lie groups
  general Linear groups 106
    orthogonal groups 106
      generalized orthogonal group 106
      Lorentz group 116
    symplectic group 106
    unitary group 107
  special linear groups 106
    special orthogonal group 106
    special unitary groups 107
linear operator 31, 96
linearly independent set 30
  linear span of 30
Lorentz transformation 139

mesons 116
Mulliken symbols 41

negative roots 109
Noether theorem 117
non-abelian 3
norm 32
normal mode 73
normal subgroup 5

one-parameter group 88
orthogonal 33

particle physics 130
  antiquarks 136
  force carriers 130
    gluons 130
    photons 130
    vector bosons 130
  hadrons 130
    baryons 130
    mesons 130, 137
  leptons 130
  quarks 132
permutation cycles 11
  two-cycle, transposition 11
permutation groups 10

# Index

Poincare algebra 117
Poincare group 117, 143
point groups 9, 21
Poisson bracket 92
positive roots 112
projection operator 47

quadrupole moment 66, 68
quark model 116, 130, 132
quaternion group 6

rank 107
reducible representation 38
   completely reducible 39
      multiplicity of irreducible components 45–46
regular representation 44–45
root vectors 107
rotational symmetry 118
   angular momentum conservation 118

scalar multiplication 30
scaling transformation 32
Schoenflies notation 23
secular equation 73
   secular determinant 73
selection rules 64, 116
self-adjoint operator 34, 60
semi-direct product 15
semi-simple Lie algebras 100
similarity transformation 32
simple group 5
simple Lie algebras 103
simple roots 109
simply connected 101
special unitary group 101
stereographic projections 23
structure constant 92, 105
subalgebra 95
   invariant subalgebra 95
subgroups 4
subrepresentation 38
symmetric group 9–10

permutation group 10
symmetric subspace 51
symmetry
   axis of 18
      equivalent 21
      principal 19
   continuous 87
   inversion 19
   plane of 19
      equivalent 21
   point of 19
   roto-reflection 19
   transformation 18

tensor product 45
   of irreducible representations 46
time reversal 67
   antilinear property 68
translational symmetry 117
   space translation 117
      linear momentum conservation 117
   time translation 118
      Hamiltonian 118

ungerade 42
unitary group 107
unitary operator 35

vector space
   complex 30
   real 36
vibrational degrees of freedom 71

weight vectors 110
Wigner–Eckart Theorem 116, 128

Young diagrams 13
   dimension of irreducible representation 121
   Young tableau method 116, 119
      symmetric group 119